# NEW SCHOOL SERIES

GENERAL EDITOR. R. Stone, M.A., A.Inst.P.
*Formerly Second Master, Manchester Grammar School*

**The Elements of Geography in Colour**

F. R. Dobson, B.A.
H. E. Virgo, M.A.

**A New School Geography**

Volume 1    THE ELEMENTS OF GEOGRAPHY
Volume 2    THE BRITISH ISLES
Volume 4    NORTH WEST EUROPE
Volume 5    CANADA AND THE UNITED STATES

**Map Reading and Local Studies in Colour**

A. P. Fullagar, B.A.
H. E. Virgo, M.A.

**A New Geology**

M. Bradshaw, M.A.

**A New Chemistry**

S. Clynes, B.Sc., F.R.I.C
D. J. W. Williams, M.A.
J. S. Clarke, B.Sc., M.A., F.R.I.C.

**A New Biology**

K. G. Brocklehurst, M.A., M.I.Biol.
Helen Ward, B.Sc.

**Le Français d'Aujourd'Hui**

Parts One, Two, Three and Four [GCE]
P. J. Downes, M.A.
E. A. Griffith, B.A.
Part Four [CSE]
P. B. Houldsworth, B.A.(Manc.), Doct. de l'Univ.(Paris)

**Starting Points**

A New English Book
G. P. Fox, M.A.
B. A. Phythian, M.A., B.Litt.

**Listen**

An Anthology of Dramatic Monologues
A. Thompson, M.A.

**Storylines**

A Teaching Anthology of Short Stories
A. Thompson, M.A.

**Touchstones**

A Teaching Anthology of Poetry in Five Volumes
M. G. Benton, M.A.
P. Benton, M.A.

# A NEW SCHOOL GEOGRAPHY Volume 4
# NORTH WEST EUROPE

**F. R. DOBSON**, B.A.
*Formerly Senior Geography Master*
*The Forest Grammar School*
*Winnersh, Berkshire*

**H. E. VIRGO**, M.A.
*Formerly Vice-Principal*
*The John Cleveland College*
*Hinckley, Leicestershire*

*Revised by*
G. N. Minshull, M.A., M.Sc.

HODDER AND STOUGHTON
LONDON SYDNEY AUCKLAND TORONTO

British Library Cataloguing in Publication Data

Dobson, Frank Ronald
   A new school geography. — 3rd ed.
   Vol. 4: North West Europe
   1. Geography
   I. Title   II. Virgo, Hugh Edward
   III. Minshull, Gordon Neil
   910      G128

ISBN 0 340 25583 8

First printed 1969
Reprinted 1971
Metric edition 1972, reprinted 1974
Paperback edition 1975
Second edition 1977
Third edition 1980, reprinted 1982

Printed and bound in Great Britain for
Hodder and Stoughton Educational,
a division of Hodder and Stoughton Ltd,
Mill Road, Dunton Green, Sevenoaks, Kent,
at the University Press, Cambridge.

# CONTENTS

# EDITOR'S INTRODUCTION

We live in a rapidly changing world: some might say in a too rapidly changing one. Today ideas can alter as much in a decade as in the whole of the preceding century. As the kaleidoscopic patterns of syllabus and method pass before us, it is not always easy to distinguish the important and permanent from the ephemeral. It is our hope that, in the New School Series, we shall be able to produce a group of books which will help colleagues in the classroom and their pupils to meet the challenge of the next ten years.

Our main purpose is to provide a wide range of books, covering both the Arts and the Sciences at a number of levels, which will allow the teacher increased latitude in his approach and offer him full scope to develop the creative aspects of his work. It must not be forgotten that the suitability of a particular text for a given form depends largely on the way in which the teacher uses it: the speed and depth of the work will be dictated by the abilities and interests of the pupils rather than by the wishes of the teacher or the nature of the text. Many of the new books will be suitable for all but the lower streams of the comprehensive schools: others will express the newer conceptual, as opposed to factual, approach to teaching but may be contained within somewhat narrower academic boundaries. We shall be greatly in debt to the many teachers, professional bodies, and others whose untiring efforts have done so much to change the pattern of teaching in recent years. Nor must we forget those pupils who, whether consciously or not, have been the guinea-pigs in the experiments which were necessary to prove the new ideas.

We hope to arrange for books to be written by teams of two or more experienced teachers who have tried out the new methods and syllabuses in the classroom, and who will be able to engender in their readers the same enthusiasm which they have instilled into their pupils. It is this dual interest of those who teach and those who are taught which is the key to all successful learning, a process in which we hope to play our part.

R. STONE

# PREFACE

North West Europe covers the physical and human geography of Scandinavia, the Benelux countries, France, East and West Germany, Switzerland and Austria, and is designed for fourth and fifth year students in secondary schools. The aim, as in previous volumes in the series, is to present the subject matter concisely and systematically to assist the pupil in learning and understanding.

Since the early 1970's, the geography of North West Europe has been particularly affected by several significant phenomena. These include dramatic changes in energy supply and as a consequence the depressed economic outlook. Another is the existence, consolidation and enlargement of the European Economic Community as a powerful economic grouping. These phenomena have had a substantial influence upon the geography of energy, agriculture, industry, population and city development. In this edition, therefore, population statistics, diagrams and production figures have been revised and brought up-to-date to take account of these rapid and substantial changes.

Appreciation must be expressed to the Norwegian, Swedish, Danish, Dutch, Belgian, French, German, Swiss and Austrian embassies for supplying information and photographs from their countries. Thanks are also due to the Swedish National Tourist Office, Swedish Iron Ore Company, Norwegian Tourist Office, Belgian National Tourist Office, Institute Belge, Documentation Française, Swiss National Tourist Office, and Austrian State Tourist Department, Aerofilms and J. Allen Cash for their considerable help with photographs, and to the editor, Mr. R. Stone, for his useful suggestions.

G. N. MINSHULL

# 1: INTRODUCTION – PHYSICAL GEOGRAPHY

## The Building of the Land

Although North-West Europe covers only a small part of the globe, its physical features are of great variety, ranging from rugged Alpine ranges and high plateaus to rolling hills and flat plains. The coastal forms too are very varied, with numerous peninsulas and inlets which allow the influence of the sea to penetrate far inland.

All these features are the result of a long and complicated geological history, with quiet periods of deposition during which thousands of metres of limestone, sandstone, clay and other sedimentary rocks were laid down, mainly on the sea bed. These quiet periods were interrupted at times by violent earth movements which compressed and uplifted the newly-formed rocks into ranges of fold mountains. The movements were accompanied by large-scale faulting and igneous activity.

The oldest rocks in Europe, of Pre-Cambrian age, occur to the west and north of the Baltic Sea, where they have been worn down over millions of years to form a lowland or peneplain, to which the name Baltic Shield has been given. Subsequently this shield has acted as a resistant block against which earth movements were directed, so that fold mountains were formed round its margins.

The earliest earth movements of which there is widespread evidence took place at the end of the Silurian period, and led to the formation of the Caledonian fold mountains. These ancient mountains were worn down to peneplains, but were subsequently re-uplifted, and their remnants now form the Scandinavian Highlands, the Scottish Highlands, and other highlands in northern Britain.

Earth movements at the end of the Carboniferous period gave rise to the Hercynian fold mountains, whose remnants form a series of resistant plateaus or horsts including the French Massif Central, Armorican Peninsulas, Vosges, Black Forest, Ardennes, Middle Rhine Uplands, Harz Mountains and Bohemian Plateau.

The youngest fold mountains in Europe were formed as recently as the Miocene period, about thirty five million years ago, and are known as the Alpine fold mountains. Because of their youth, these mountains form most of the highest ranges in Europe, including the Alps, Jura and Pyrenees. So intense were the Alpine earth movements that over-

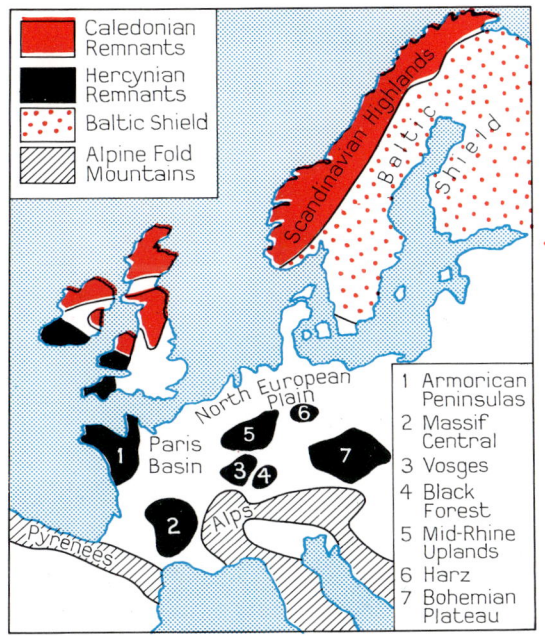

FIG. 1.1 **The structure of North West Europe**

folding occurred on a large scale, and there is evidence of nappe structure i.e. the thrusting forward of enormous folds over the top of other folds. There was also much faulting not only within the folded zone, but also in older rocks to the north, and there was widespread igneous activity with volcanic outpourings as far apart as north-west Britain, the Massif Central in France, and the Eifel in Germany.

North of the Hercynian and Alpine zones is a broad plain, stretching through northern France, Belgium, the Netherlands and north Germany, and called the North European Plain. It is built of sedimentary rocks which have not been greatly affected by mountain-building movements, although gentle folding of Alpine age did occur in southern England and the Paris Basin.

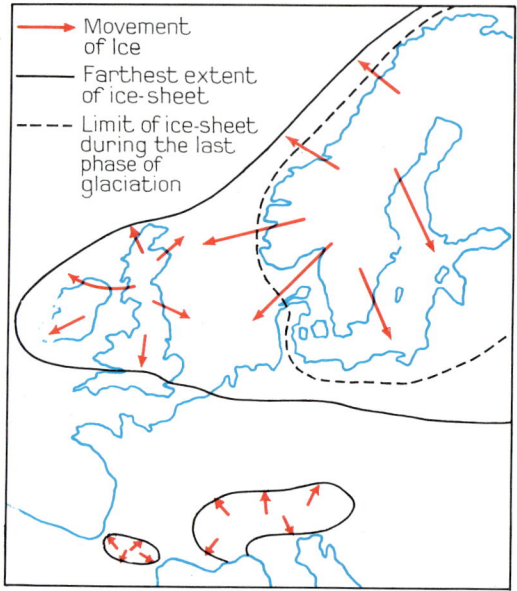

FIG. 1.2 **North West Europe in the Ice Age**

## The Ice Age

Some 600,000 years ago, in the Pleistocene period, the climate became colder and ice sheets spread out from the Baltic region, eventually covering a large part of northern Europe. Smaller centres of glaciation also developed in a number of high mountain areas, notably the Alps, Pyrenees, and the highlands of northern Britain. The Ice Age consisted of four glacial periods, separated by three inter-glacial periods when mild conditions returned. The last glacial period ended as recently as 15,000 years ago, although in the Alps, Pyrenees and Scandinavian Highlands, snow fields and glaciers remain at high levels.

## GEOLOGICAL PERIODS

| | | |
|---|---|---|
| QUATERNARY | Recent (Submergence of Continental Shelf) | |
| | Pleistocene (The Ice Age) | |
| TERTIARY | Pliocene | |
| | Miocene (Alpine Earth Movements) | |
| | Oligocene | |
| | Eocene | |
| SECONDARY | Cretaceous | |
| | Jurassic | |
| | Triassic | |
| PRIMARY | Permian | |
| |     (Hercynian Earth Movements) | |
| | Carboniferous | |
| | Devonian | |
| |     (Caledonian Earth Movements) | |
| | Silurian | |
| | Ordovician | |
| | Cambrian | |
| | Pre-Cambrian (Rocks of Baltic Shield) | |

Erosion and deposition by the ice had a marked effect on the surface features and soils of many European countries. Erosion has left its mark mainly in mountain regions, as shown by the pyramid-shaped peaks, narrow sharp-edged ridges (arêtes), armchair-shaped hollows (cirques), and deep U-shaped valleys, bordered by hanging valleys and truncated spurs. In lowland areas the effects of deposition are more apparent, with lines of low hills formed by the terminal moraines, extensive deposits of boulder clay from the ground moraines, and sheets of sandy outwash material left by melt-water streams flowing from the edge of the ice sheets. Fine dust, carried by winds from the exposed ground moraines at the end of the Ice Age, settled over wide areas to form superficial deposits of loess.

During and since the Ice Age there have been considerable fluctuations in the levels of land and sea. The enormous weight of the ice pressed the land down, and when the ice melted huge quantities of water were added to the oceans. The rise in sea level led to the drowning of many lowlands including the areas now occupied by the

North Sea and English Channel and the straits connecting the Baltic Sea and North Sea. The submergence of river valleys produced a highly indented coastline, with many peninsulas, estuaries, rias and fiords. An opposite movement also took place when the land, relieved of its weight of ice, started to rise again. Evidence of this rise in land level can be seen in the numerous raised beaches in Scotland, Scandinavia and other parts of North-West Europe.

### The Climate of North-West Europe

In order to form a general picture of the climate of North-West Europe, various aspects including temperature, pressure, winds and precipitation, must be considered.

FIG. I.3 **January isotherms**     FIG. I.4 **July isotherms**

**Temperature.** Figures 1.3 and 1.4 show the mean temperatures for January and July, and it will be seen that in January the isotherms run roughly from north to south, whilst in July they run from north-east to south-west. The chief factors determining these temperature distributions are:

(a) *Differences of Latitude*. The amount of heat received from the sun at any place depends on the angle of the sun's rays and the length of daylight, both of which vary with latitude. As a result, places in a southerly latitude receive more heat than those farther north.

(b) *Distance from the Sea*. The Atlantic Ocean, and to a lesser extent the Mediterranean Sea and Arctic Ocean, have important effects on land temperatures. In summer the sea is cooler than the land and in winter it is warmer. It therefore has a moderating effect on the temperatures of coastal areas which are said to have a maritime climate, in contrast to interior areas which have a more continental climate, with greater extremes of temperature. The influence of the sea is most marked in the case of the Atlantic Ocean, which is warm for its latitude because of the effect of the North Atlantic Drift, a warm ocean current originating in the Gulf of Mexico.

(c) *Winds*. Europe has no north-south mountain barriers, and winds are able to blow right across the continent. Maritime influences are carried eastwards by the prevailing westerly winds, but sometimes in winter, cold easterly winds may bring sub-zero temperatures from eastern Europe to much of western Europe.

(d) *Altitude*. The temperature falls, on average, by $1°C$ per 150 metres of ascent. It must be remembered that isotherms represent the temperatures at sea level, and that to find the actual temperature at any point in figures 1.3 or 1.4, it is necessary to subtract $1°C$ from the isotherm value for every 150 metres it is above sea level.

**Pressure, Winds and Precipitation.** In the North Atlantic Ocean south-westerly winds blow from the Azores High Pressure belt to the Icelandic Low Pressure belt, and much of north-west Europe comes under their influence. The south-westerly winds contain warm, moist air of tropical origin, and this air meets cold air of Arctic origin around latitude 60 degrees north. The surface of contact is called the Polar Front, and the conflict between the warm and cold air gives rise to depressions which travel eastwards, bringing changeable weather with rain to much of north-west Europe.

In winter the Icelandic Low is well-developed whilst pressure over eastern Europe is high. The south-westerly winds, accompanied by depressions, bring mild, rainy weather to western Europe and also to the Mediterranean lands. The weather in central and eastern Europe tends to be anti-cylonic, with high pressure, intense cold and little precipitation. Sometimes in January and February these anti-cyclonic conditions extend westwards, bringing severe weather to western Europe.

In summer the Icelandic Low is less powerful and the Azores High extends northwards to cover much of the Mediterranean Sea. Pressure over eastern Europe is now relatively low, so that depressions follow a more northerly route and travel far into central and eastern Europe.

Thus there is rainfall both in north-west Europe and in central and eastern Europe, whilst the Mediterranean lands have dry, sunny weather broken only by occasional thunderstorms.

Another important factor governing the total annual rainfall is the height of the land. The importance of relief rain can be seen in the close relationship between the areas of heaviest rainfall and the highest land.

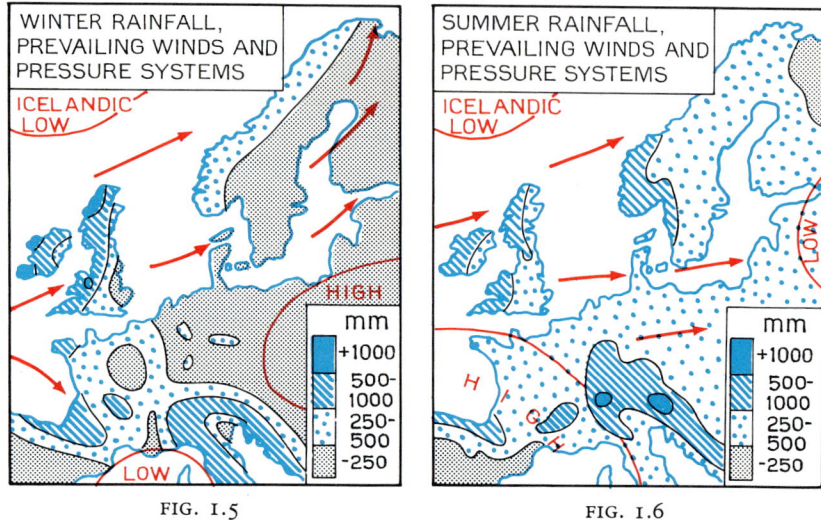

FIG. 1.5

FIG. 1.6

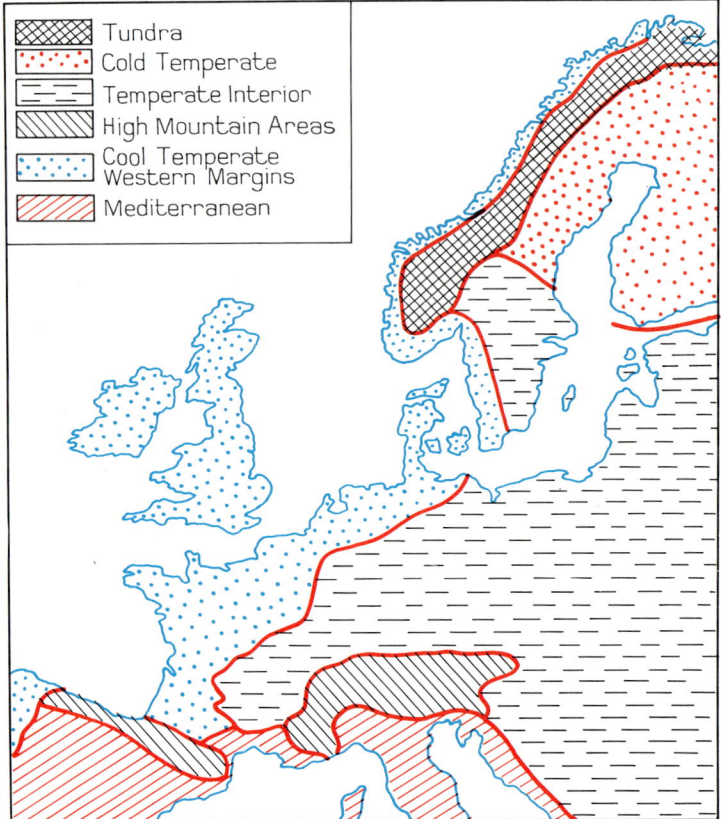

FIG. 1.7 **Climatic regions**

## Climate and Vegetation Regions

*Cool Temperate Western Margins.* The western and southern coastlands of Norway and Sweden, the whole of Denmark, north-west Germany, the Netherlands, Belgium and western France belong to this region. Being fully exposed to Atlantic influences, they have a maritime climate, with mild winters, cool summers, and rain throughout the year.

The mild, damp conditions favour the growth of broad-leaved deciduous forest, which includes oak, elm, ash, beech and other valuable hardwoods. On high ground and poor soils there is heath vegetation of heather, gorse, bracken and coarse grasses, with scattered silver birch and pine trees. Much of the land has now been cleared for farming and human settlement.

*Temperate Interior.* This region includes large parts of eastern France and Germany, as well as Switzerland and Austria north of the Alps.

Compared with the western margins the influence of the ocean is less marked and the climate is more continental, with winters several degrees colder and summers slightly warmer, in comparable latitudes. Most of the rain falls in spring and summer, when pressure is low over eastern Europe and the westerlies are able to penetrate the region. There is light snowfall in winter.

The natural vegetation is mixed forest, with mainly deciduous trees on the lower land, and conifers on the hill slopes.

*Mediterranean Region.* The only part of North-West Europe belonging to this region is southern France. Its main characteristics are hot, dry summers and mild, wet winters, and the natural vegetation of small, evergreen trees and shrubs with deep root systems is adapted

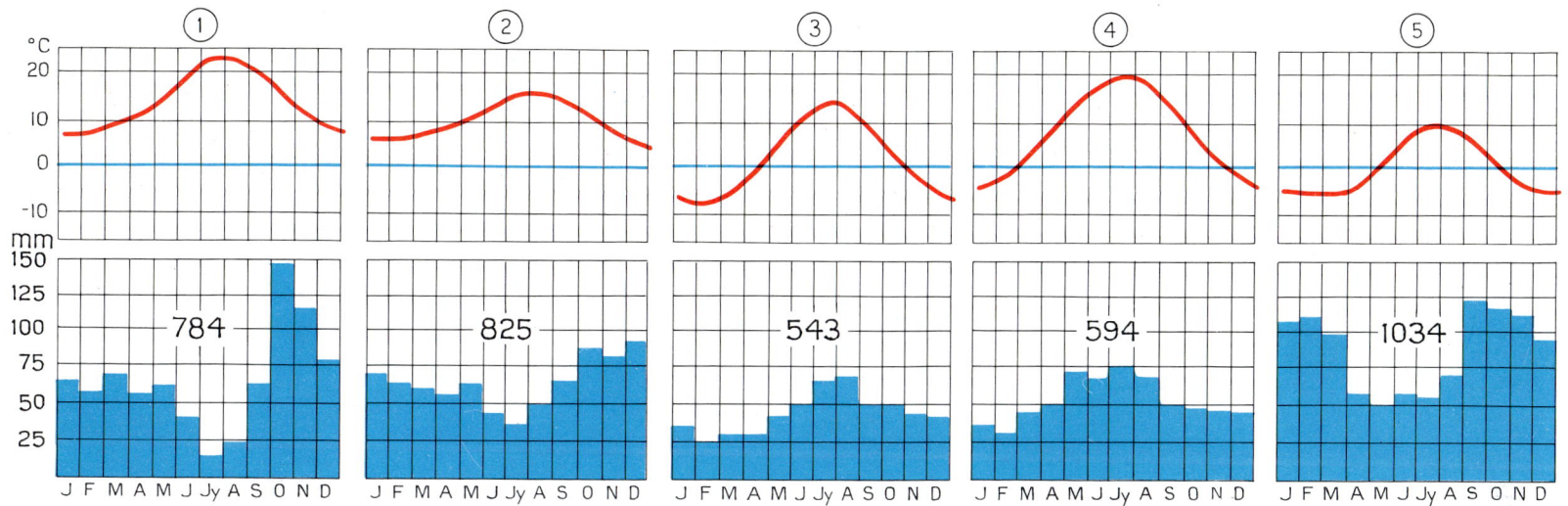

FIG. I.8 **Temperature and rainfall graphs for five selected stations**

to withstand the long summer drought. The wild olive, Mediterranean pine and cypress are typical trees, but much of the land is covered by a dense tangle of low shrubs called maquis.

*Cold Temperate Region.* Because of its northerly latitude, north-east Sweden has a cold temperate climate. Winters last longer and are more severe, and summers are cooler, than farther south. It has a low precipitation, but also a low rate of evaporation, so that there is sufficient soil moisture for tree growth. Much of the land is covered with coniferous forest, including pine, spruce and fir which, being evergreens, can make full use of the short growing season. The trees are mainly softwoods, and there is a large demand for the timber.

*Tundra.* Much of the northern and interior parts of Scandinavia are classed as Tundra. Winters last eight or nine months, during which time the ground is frozen and snow-covered and there is little daylight. During the short summer the land surface thaws, but the subsoil remains frozen. There is a variety of vegetation, consisting of mosses, lichens, grasses, flowering plants, and hardy shrubs such as the cranberry and bilberry.

*High Mountain Areas.* In high mountains the climate and natural vegetation vary with the altitude. In general the temperatures are lower and precipitation higher than at lower levels. Distinct vegetation belts can be recognised. There are mountain forests consisting

mainly of coniferous trees, but sometimes with a lower belt of deciduous trees; above the forests are Alpine meadows which are snow-covered in winter. The Alps, Pyrenees and Scandinavian Highlands rise above the snow line, which lies at about 900 metres in northern Norway and 2,700 metres in the Alps.

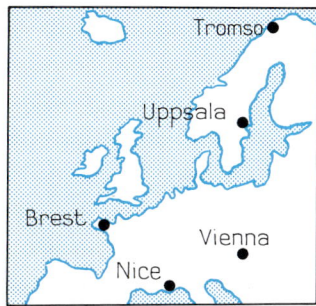

FIG. I.9 **Map of five weather stations**

## Exercise

Pair each of the above climate graphs with one of the towns on the map, and state why you have done so. Describe the natural vegetation you would expect to find in the vicinity of each town.

# 2: INTRODUCTION – HUMAN AND ECONOMIC GEOGRAPHY

North-West Europe contains some of the most densely populated and highly industrialised countries in the world. France, Germany, the Netherlands and Belgium were great colonial powers and became important trading nations, building up economies which were based on imports of raw materials and foodstuffs, and exports of manufactured goods. Norway, Sweden, Denmark, Switzerland and Austria also depend to a considerable extent on foreign trade for their prosperity.

The great devastation and loss of life during the two world wars of the twentieth century have been serious setbacks, but recovery has been fast, and the last twenty-five years have seen great industrial progress. As a result of the 1939–1945 war, Germany lost large territories to Poland and the Soviet Union, and has emerged as two separate states, the German Federal Republic (West Germany) and the German Democratic Republic (East Germany). Partly as a result of the wars, and partly because of the prosperity of west European countries, several million immigrants have come into the region, particularly into West Germany, France and Switzerland, from southern and eastern Europe. At the same time there have been considerable population movements within each country. There has been a drift of people from the countryside to the towns, due to the attractions of town life, and also because mechanised farming requires fewer people to work on the land. Another phenomenon has been the growth of great residential suburbs which sprawl out for many miles on the outskirts of the major cities.

Originally in 1952 six countries – West Germany, France, Italy, Belgium, the Netherlands and Luxembourg – formed the European Coal and Steel Community with the object of pooling their coal and iron resources and developing their steel industries. In 1958 these countries formed the European Economic Community (E.E.C.) with the aim of creating a single economic unit by means of the abolition of tariffs, and common economic policies between all its member-states.

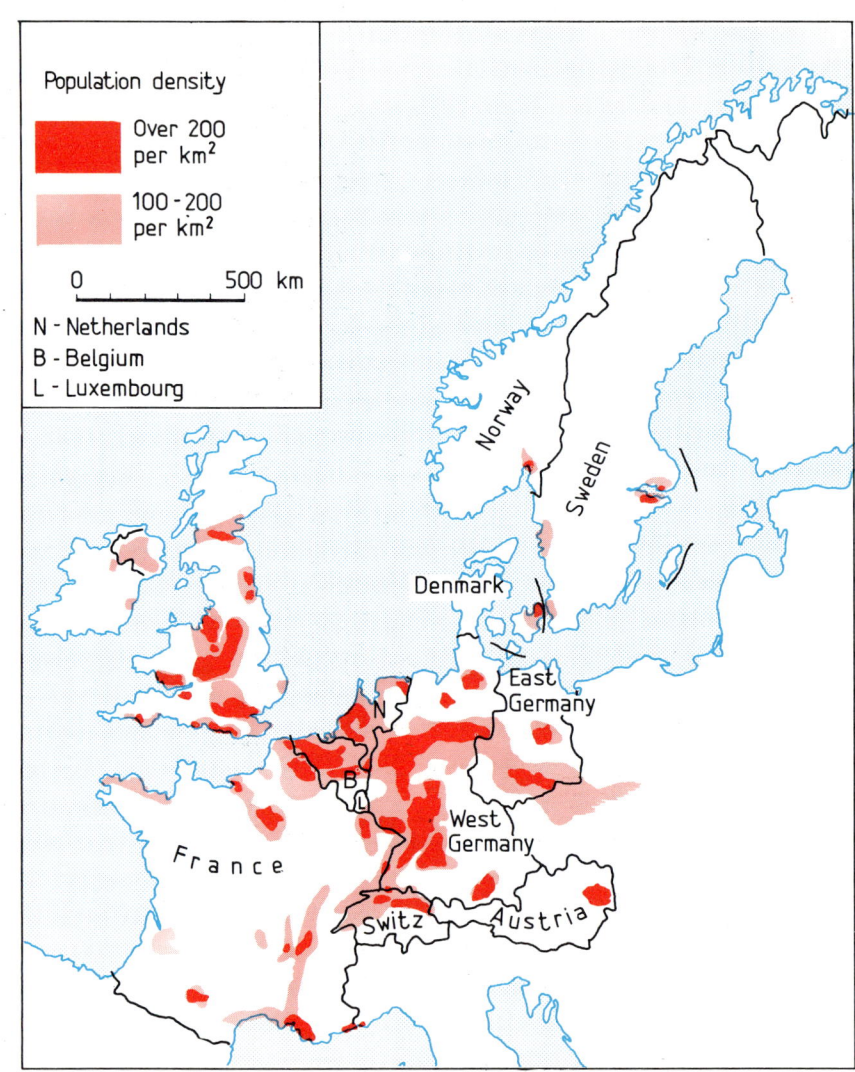

FIG. 2.1 **Density of population**

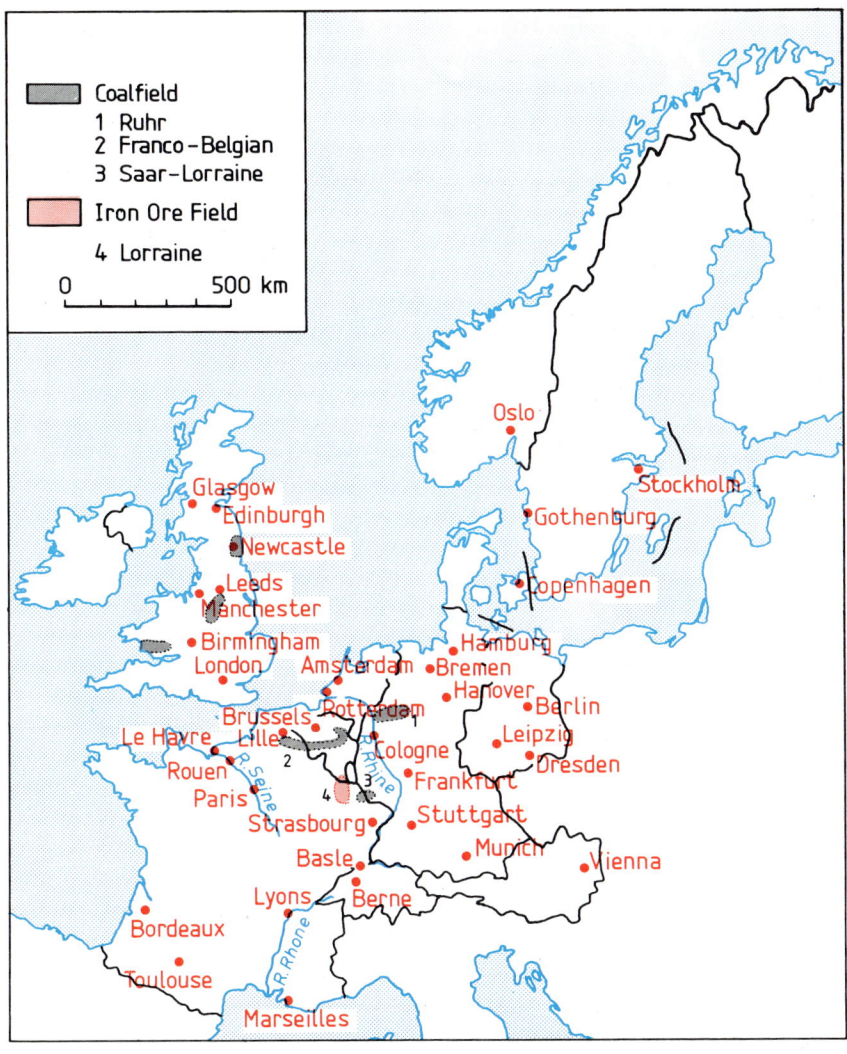

**FIG. 2.2 Major cities and mineral deposits**

Using both diagrams 2.1 and 2.2, describe the distribution of the most densely populated parts of North-West Europe. What reasons can you give for this distribution? Figures 1.7 and 2.4 should provide further information.

The success of the E.E.C. in overcoming old prejudices and in helping to create greater wealth and to forge strong trading, economic and cultural links throughout the 1960's, has been accompanied by a considerable expansion of its membership. In 1973, Denmark, Eire and Great Britain acceded to the E.E.C. Greece will be admitted in 1981 and Spain and Portugal are negotiating to join. In the foreseeable future, the E.E.C. will be enlarged to 12 member-states, a very powerful economic unit on a par with the United States and U.S.S.R.

## Agriculture

North-West Europe is one of the most productive agricultural regions in the world, but so great is its population that large quantities of foodstuffs have to be imported from overseas. There is also a considerable trade in specialised farm products between the individual countries. Owing to the high level of mechanisation, under ten per cent of the working populations of most North-West European countries is employed in agriculture. Mixed farming is generally practised, much of it of an intensive type whereby high yields are obtained from small units of land. The character of the farming is influenced by a number of factors, including (a) the generally maritime climate with well-distributed rainfall, (b) the large industrial concentrations making a demand for locally-produced vegetables, fruit and dairy products, and (c) the great variety of soils, ranging from fertile peats, silts and loess, to sterile sands, and thin, acid mountain soils.

Just as there are great differences in the physical geography, there are also differences in the type of farming in various parts of the region, as follows:

(1) Areas where little or no farming takes place include the Tundra, and the high mountain regions of Scandinavia, the Alps and Pyrenees.

(2) A northern farming zone stretches round the south-west and south of Norway into central Sweden. Here the chief crops are oats, potatoes and hay, and dairying is the most important aspect of stock farming.

(3) On the North European Plain in northern France, the Low Countries, north Germany, Denmark and south Sweden, there is an emphasis on livestock farming, especially dairying and the raising of beef cattle, and also the keeping of pigs, sheep and poultry. On the lighter soils, including the loess, arable farming is well represented with large acreages of wheat, barley, potatoes and sugar beet, and important market garden concentrations.

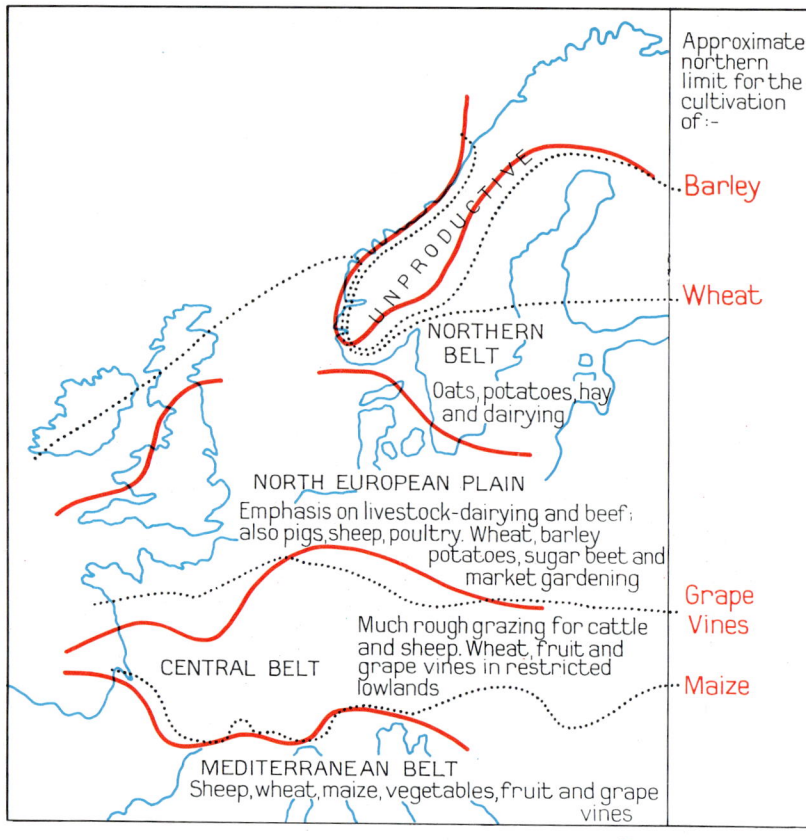

FIG. 2.3 **Agricultural belts**

Within the figure:

Approximate northern limit for the cultivation of:—

Barley

Wheat

UNPRODUCTIVE

NORTHERN BELT
Oats, potatoes, hay and dairying

NORTH EUROPEAN PLAIN
Emphasis on livestock—dairying and beef; also pigs, sheep, poultry. Wheat, barley potatoes, sugar beet and market gardening

Grape Vines

CENTRAL BELT
Much rough grazing for cattle and sheep. Wheat, fruit and grape vines in restricted lowlands

Maize

MEDITERRANEAN BELT
Sheep, wheat, maize, vegetables, fruit and grape vines

(4) Across central France, the Belgian Ardennes, south Germany, and parts of Switzerland and Austria, the general character of farming is influenced by the large proportion of poor upland soils, with rough grazing for cattle and sheep. In contrast, there are favourable conditions for the growing of wheat, fruit and grape vines in some sheltered valleys.

(5) In southern France the hot, dry summers show the influence of the Mediterranean climate. Here there are fewer cattle, and sheep are the chief livestock. Wheat is still important, but maize and even rice are also grown. Irrigation is increasingly relied on especially for the cultivation of vegetables, and again fruit and grape vines are important.

*Fishing*

The waters off the coasts of North-West Europe are rich in fish, especially the shallow waters of the continental shelf including the North Sea. The main reason for this is the abundance of plankton, the minute organisms which are the chief food of fish. The indented coast line provides numerous sheltered harbours for fishing craft. At the same time the limited resources of some coastal areas have led many communities to turn to the sea for a living, as in Norway and Brittany. The densely populated industrial regions of western Europe provide the market for most of the fish. The leading fishing nation of the region is Norway, with West Germany, France and Denmark all of major importance.

*Industry*

The Industrial Revolution led to a concentration of industry on the coalfields, and even today the two largest, the Ruhr and Franco-Belgian coalfields, form the major industrial regions of North-West Europe. Nevertheless during the twentieth century the distribution of manufacturing industries has become more widespread partly due to improved communications and partly to the increasing use of forms of energy other than those provided by solid fuels. In many areas, notably in Norway, Sweden, southern France, Switzerland and Austria, it has been possible to base industry on hydro-electricity. Even more recently mineral oil, which is mainly imported, and natural gas, which is produced in the Netherlands, West Germany and south-west France, have begun to influence industrial development.

Although many North-West European countries have to import large quantities of raw materials for their industries, there are considerable local resources including farm produce, hardwood and softwood timber, and minerals. The region is fortunate in having two major iron ore deposits, one in north Sweden and the other in eastern France (Lorraine), as well as smaller ones of local importance. Other significant mineral deposits are the bauxite of south-east France, and the potash of eastern France, East Germany and West Germany.

In many areas traditional skills and crafts have been passed on from generation to generation, especially in textile and clothing manu-

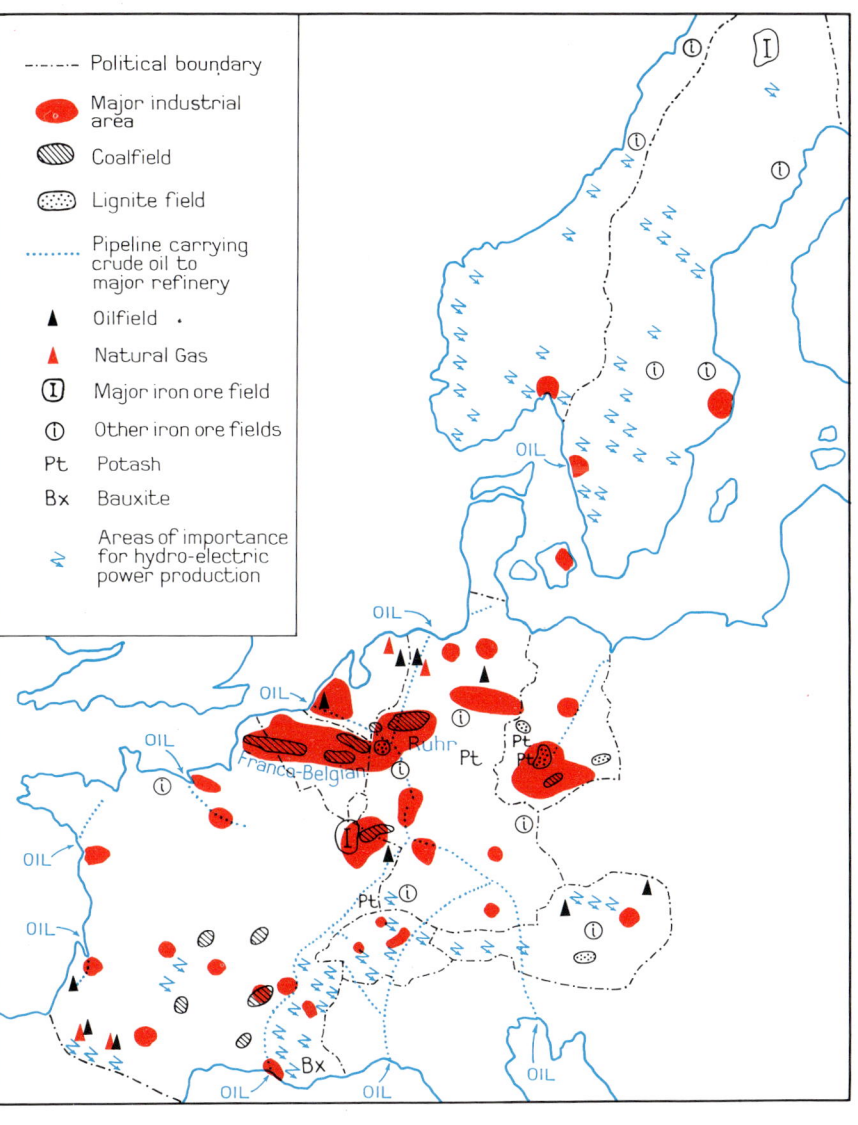

**FIG. 2.4 Power and industry in North West Europe**

Legend:
- ─··─ Political boundary
- Major industrial area
- Coalfield
- Lignite field
- ········ Pipeline carrying crude oil to major refinery
- ▲ Oilfield
- ▲ Natural Gas
- Ⓘ Major iron ore field
- ① Other iron ore fields
- Pt Potash
- Bx Bauxite
- Areas of importance for hydro-electric power production

facture, and in leather and metal-working, and during the Industrial Revolution these skills played an important part in developing the new techniques, many of which were invented and perfected in North-West Europe.

Most of the countries are maritime powers, with a long history of trading, and large merchant fleets. Since they rely on imports of raw materials to maintain their industries, and overseas markets for the finished goods, this trading experience has proved a great advantage. In many cases economic links have been maintained with former colonies, the latter continuing their traditional functions as suppliers of raw materials, and receiving in exchange, help and equipment for setting up new industries, as well as consumer goods.

The prosperity of North-West Europe, although mainly a product of geographical and historical factors, also owes much to the efforts of outstanding individuals, for example Krupp (Germany), Philips (Netherlands) and Michelin (France) who have created vast enterprises whose products are known throughout the world.

### Types of Industry

There are four main groups of industry associated with the major industrial areas. These are textiles, iron and steel, engineering and chemicals, and the factors which have influenced the location of each of them show significant variations.

The present distribution of the *textile industry* still shows the influence of its distribution before the Industrial Revolution. For example, the textile industry of north Belgium (Flanders) perpetuates the mediaeval crafts which were carried on there and had such a great reputation. The early stages of the Industrial Revolution brought a concentration of textile manufacturing to the coalfields, especially those of northern France and the Ruhr, since coal was needed as fuel for the steam-powered machinery. Since then, the increasing use of synthetic fibres has provided a link with the chemical industry, and at the same time the use of electricity as the source of power has meant that new factories have a much more scattered distribution.

The location of the *iron and steel industry* has been influenced by proximity to the coalfields and iron ore deposits, and the need for easy assembly of raw materials and finished goods. The greatest concentrations of the industry are still on the Ruhr and Franco-Belgian coalfields, and on the Lorraine ore field in eastern France. More

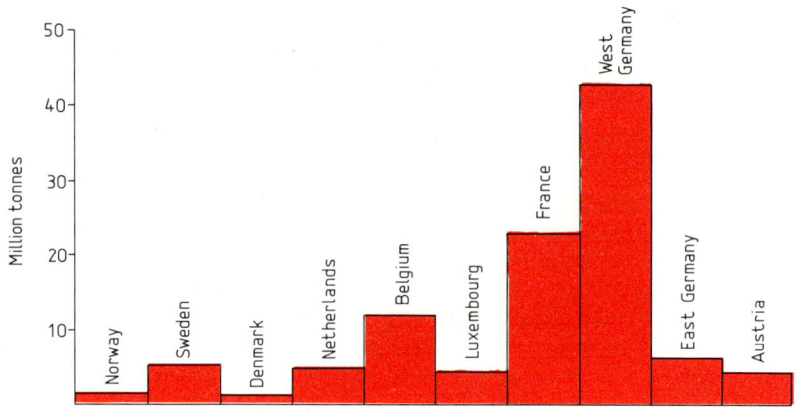

FIG. 2.5 **Crude steel production**
(Average annual production 1977)

recently-established plants in northern France, the Netherlands, Germany, Austria and Norway have been influenced by the advantages of a coastal location, or by the availability of hydro-electricity.

*Heavy engineering*, which is concerned with the production of steel beams for buildings, boilers and other bulky articles, is closely linked to the iron and steel centres because of the large quantities of metal required. *Light engineering*, which includes electrical engineering and the production of machinery, vehicles and aircraft, is very scattered and is well-represented in most settled parts of North-West Europe, since the costs of bringing in the raw materials and distributing the finished articles make up only a small proportion of the total costs.

Traditional centres for the *chemical industry* have been the Franco-Belgian and Ruhr coalfields, but the enormous expansion of the industry during the last twenty-five years has been associated with coastal sites particularly near to oil refineries which are a source of raw material especially for plastics and synthetic fibres. The Rotterdam and Marseilles areas are outstanding examples. However, as in the case of engineering, branches of the chemical industry are now represented in many manufacturing towns.

## Exercises

1. What general factors account for North-West Europe being one of the most densely populated parts of the world?
2. What does E.E.C. stand for, and what are its aims?
3. Explain why mixed farming is so widespread in North-West Europe.
4. Account for the importance of fishing on the European continental shelf.
5. Why are so many manufacturing industries located on the coalfields?

# 3: NORWAY

Much of Norway is built of very old rocks which were intensely folded during the Caledonian earth movements, then worn down throughout millions of years, and later uplifted to form the present high plateau known as the Scandinavian Highlands. The plateau surface has been cut into by rivers, and during the Ice Age it was heavily glaciated. As a result, the valleys were deepened, and along the west coast their lower portions were submerged to form fiords. In the south-east the land is lower, with broken relief due to faulting, and there are extensive glacial deposits.

The dominant south-westerly winds and the North Atlantic Drift are the major influences on Norway's climate. Consequently the summers are cool, and the winters are so mild that the whole coastline remains ice-free, although on the interior uplands conditions are more severe. Exposed west coast areas have a heavy precipitation (for example Bergen has 1,854 mm per annum), with much snow inland, whilst the sheltered south-east lowlands are considerably drier (Oslo has only 737 m).

Norway has a population of over 4 million and covers 325,000 square kilometres. The density of population (12 per square km) is the lowest in Europe, and most of the people live on the south-east lowlands and along the narrow coastal plains. The high proportion of barren plateau and the northerly latitude of much of the country account for the almost complete lack of population in the interior and north.

### Agriculture

Only 3 per cent of the land is farmed. Farming is particularly concerned with livestock, and much of the arable land is used to produce fodder crops for cattle, pigs and sheep. Oats and barley are the chief cereals, since it it generally too wet to produce much wheat. Farms are very small and the farmer often combines agriculture with another occupation such as fishing, lumbering or even factory-work. There is a marked trend towards fewer people working on the land. Holdings are being combined so that machinery can be used more effectively. This has led to greater efficiency, and production is increasing in spite of a decline in the number of farm workers.

### Fishing

Norway is one of Europe's leading fishing nations. The abundance of plankton on the continental shelf accounts for the rich and varied

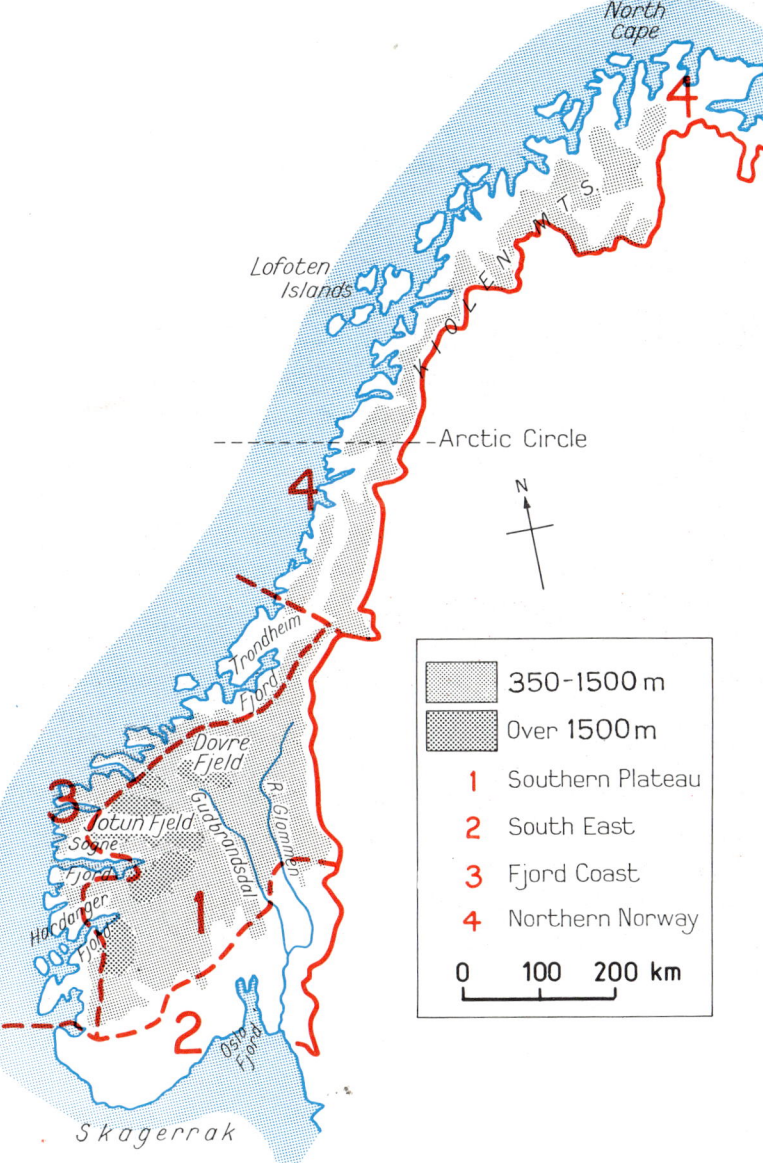

FIG. 3.1 **The regions of Norway**

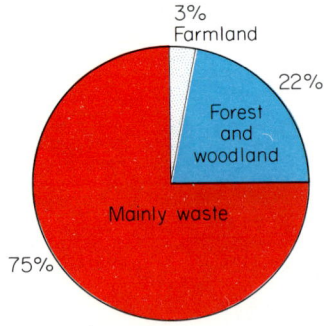

FIG. 3.2 **Norway—land use**

fish life, and the sheltered, ice-free fiords provide excellent sites for harbours. The lack of other resources has, in the past, compelled many Norwegians to look to the sea for a livelihood, but fishing has become somewhat less important as new opportunities for employment in manufacturing industries have been created. In general cod is the chief catch in the north followed by haddock and halibut, whilst herring, mackerel, brisling and sprats are caught in the south, although the numbers of herring and sprats has decreased considerably in recent years. The industry is still poorly organised, partly because many are only part-time fishermen, and most fishing boats are small, open craft, but there are larger vessels which fish in Icelandic and Greenland waters, and also whaling fleets which operate in Antarctic waters. Fishing is the basis for many associated activities, including an expanding frozen-fish industry as well as canning and the production of fish oils and fertilisers.

### Lumbering

Forest or woodland cover 22 per cent of the land surface, but much of the finest timber has already been cut and the industry is of much less importance than in Sweden. In order to preserve the existing forest cover large scale re-planting is taking place in many areas. Pine and spruce are the commonest trees and three-quarters of the present output is from the south-east, particularly along the Glommen valley. The main uses of the timber are in the pulp and paper, furniture and match industries, and in house-building.

### Industry

Manufacturing industries have been handicapped in the past by the lack of coal, the small reserves of good quality iron ore (although

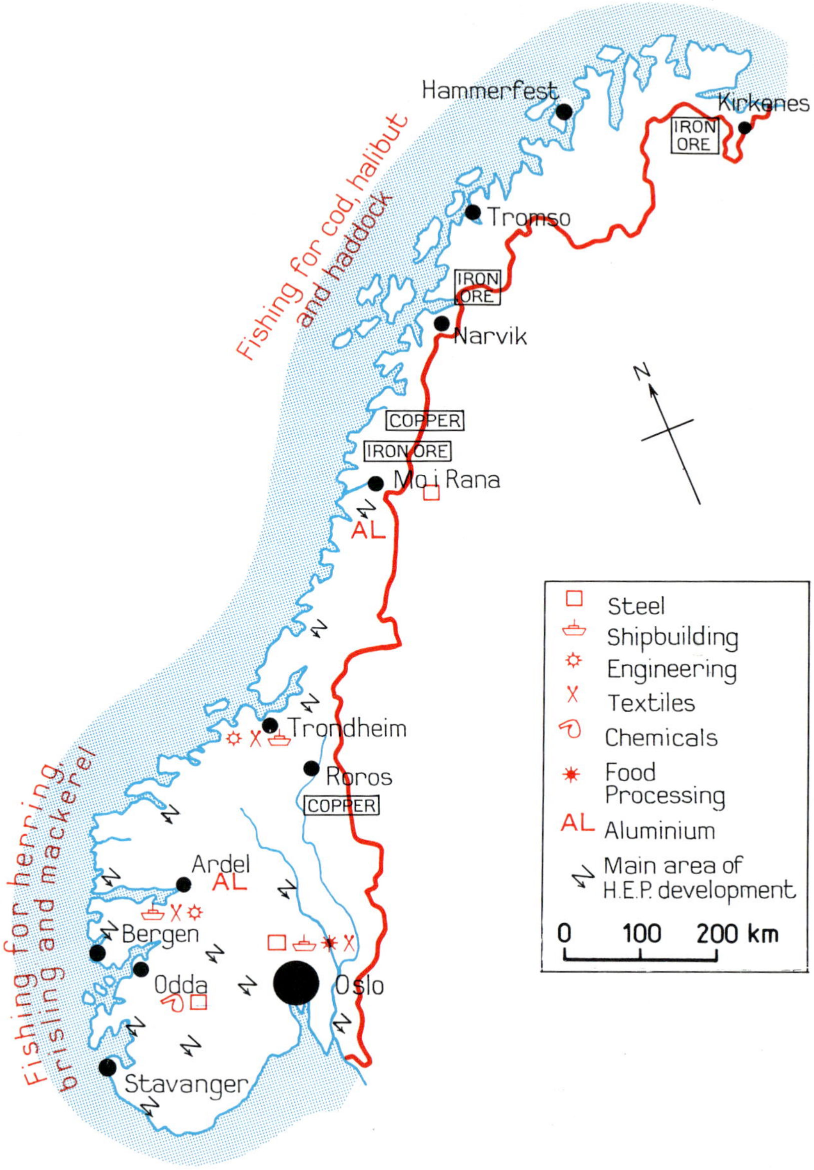

FIG. 3.3 **Norway—economic**

there is plenty of poor quality), and the smallness of the home market. However industry has developed rapidly since the war, and today some 33 per cent of the total labour force is employed in manufacturing. Norway's greatest industrial asset is an abundance of hydro-electric power, which is helped by its many lakes, waterfalls, deep, glaciated valleys and heavy rainfall. Whilst most of the power stations in the centre and north of the country supply local industries, those of the more-developed south and south-east supply an electricity grid system. So far only 20 per cent of the potential power has been harnessed, and there are plans for a great expansion of output in the future.

There are two main groups of industries. Firstly there are the long-established industries associated with the processing of primary products, including those based on timber, fish and agricultural produce, as well as clothing and textile industries. Secondly there are industries which are closely associated with hydro-electric power, including electro-metallurgical industries (the smelting and refining of aluminium, iron and steel, pyrites and copper) and electro-chemical industries, including the production of fertilisers.

Inland communications present many problems. Roads from the west coast to the interior have to negotiate steep gradients and are normally blocked by snow in winter. Railways are generally kept open with the aid of snow ploughs, and in places snow sheds are needed to protect the lines from avalanches.

Fortunately the seas are free from ice and frequent shipping services link most of the coastal settlements. Passenger ships run regular cruises carrying thousands of tourists who come each year to visit the fiords. Norway is fourth among the world's shipping nations, and its merchant fleet has many specialised craft such as oil tankers, timber ships and ore carriers. Many of these ships operate on world routes, carrying goods for other nations.

## THE REGIONS OF NORWAY

### The Southern Plateau

In the southern part of Norway the plateau is at its highest and broadest, and is known as the Fjeld. Much of it has a rounded appearance, having been smoothed by ice sheets during the Ice Age, but there are several sharp peaks rising to 2,000 and 2,500 metres in the Dovre and Jotun Fjelds, which probably remained above the ice. Even today there are a number of ice sheets and short valley glaciers.

Because of the severe climate most of the Fjeld is above the tree line

**The Norwegian Fjeld**          *Royal Norwegian Embassy*
This view was taken from Galdhopiggen, the highest mountain in Norway. Note the snowfields and the West Memuru Glacier which is in the middle of the photograph.

and has a tundra-type vegetation. On the lower land there are summer pastures (called saeters), grazed by sheep and cattle, and in some of the deeper valleys there are hydro-electric stations supplying local chemical and metallurgical industries, as at *Roros*, near the head of the Glommen valley, where low-grade copper ores are mined and smelted, and at *Lokken*, south-west of Trondheim, where sulphur pyrites are processed.

**Countryside near Oslo**

J. Allan Cash

Explain how geographical factors have influenced the land use shown in this photograph.

## The South-east

Here, round the shores of Oslo Fiord and the lower Glommen valley, is an undulating lowland covered with glacial deposits. Over half the total population live in this region, which has a drier and somewhat warmer climate than other parts of Norway. Nevertheless it is a region of contrasts, with pockets of farmland on fertile clay and alluvium interspersed with stretches of forest on the glacial gravels and outcrops of ancient rocks. Most of the country's timber is produced here, and many of the farmers are employed in forestry during the winter. Hay, oats, barley and potatoes are the chief crops, and livestock is the main basis of farming, with dairying becoming increasingly important.

It is in this region that manufacturing is most developed. There are abundant supplies of hydro-electricity from the many rivers to the north and west. Long-established industries such as timber-working, butter and cheese-making, and canning especially of brisling, flourish along with the newer chemical and metallurgical industries. The biggest industrial concern is the chemical combine of Norsk Hydro, whose main factories at *Notodden* and *Ryukan* specialise in nitrogen products particularly for fertilizers. There are pulp and paper mills at *Drammen* and many other towns along the shores of the Skagerrak. *Oslo* (480,000), the capital, chief port and leading manufacturing centre of the country, lies at the head of a submerged rift valley known as Oslo Fiord. Thus it has the advantage of a sheltered, deep-water inlet, whilst the long valleys of Glommen and Gudbrandsdal and other routes from the south-west and south-east focus on it. It has varied industries, ranging from metallurgy and shipbuilding to clothing and food processing.

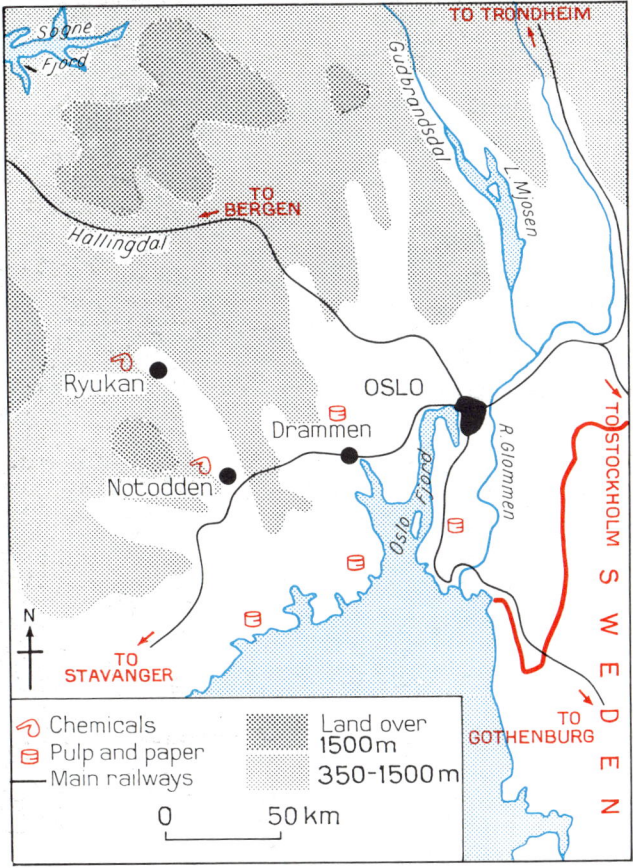

FIG. 3.4 **The position of Oslo**

**Oslo, the waterfront**                    *Royal Norwegian Embassy*

## The Fiord Coast, as far north as Trondheim

The Atlantic coast is rugged, with many fiords. These are drowned, glaciated valleys which take the form of long, narrow inlets with branches joining at right angles. There is deep water within the fiords, but a shallow 'threshold' at the entrance. The sides, which are forested in places, rise steeply up to the desolate fjeld, and there are hanging valleys where streams plunge over the edges. Along the coast is a narrow, discontinuous belt of flat land known as the strand-flat, and fringing the coast are numerous low islands called skerries which are rounded, hummocky rocks covered by glacial material.

Settlement is restricted mainly to the strand-flat and to the small alluvial plains at the head of the fiords. It is here that the dual economy of farming and fishing is practised. Because of the mild, damp climate, dairying is the main type of farming, and crops such as oats, barley and roots are grown for fodder, and grass is cut for hay. There is still some movement of cattle in summer to the saeter on the lower slopes of the fjeld, although these pastures are being increasingly given up. This seasonal movement of livestock is known as transhumance. Some of the milk is sent daily to the towns, but much is made into butter and cheese.

**Village on Sogne Fjord**

*J. Allan Cash*

Draw a simple sketch map or diagram, based on this photograph, to show the site of the village and then describe the probable occupations of the inhabitants.

**Bergen**

*J. Allan Cash*

The skerries offer protection to fishermen who fish the inshore waters within and just outside the fiords. Small, open craft are used and the catch includes cod, herring, salmon and brisling.

The broken nature of the coast together with the rugged relief of the interior means that there are few nodal points where large towns can develop. *Bergen* (215,000) is the main port and fishing centre of the region, and has shipbuilding, engineering and textile industries, as well as industries based on fish. It is linked to Oslo by railway, and lies in an area of magnificent scenery, so that it is visited by many tourists. *Trondheim* has developed in a small, fertile depression on the south side of Trondheim Fiord, at a focus of three railways, and has fishing and shipbuilding industries. Another large fishing port is *Stavanger*, in the south.

Along the coast are many hydro-electric stations and industries associated with them, for example the largest aluminium works in Norway is at *Ardel*, at the head of Sogne Fiord, and there are both chemical and metallurgical industries at *Odda*, at the head of Hardanger Fiord. It is surprising how little labour is employed in these highly automated works.

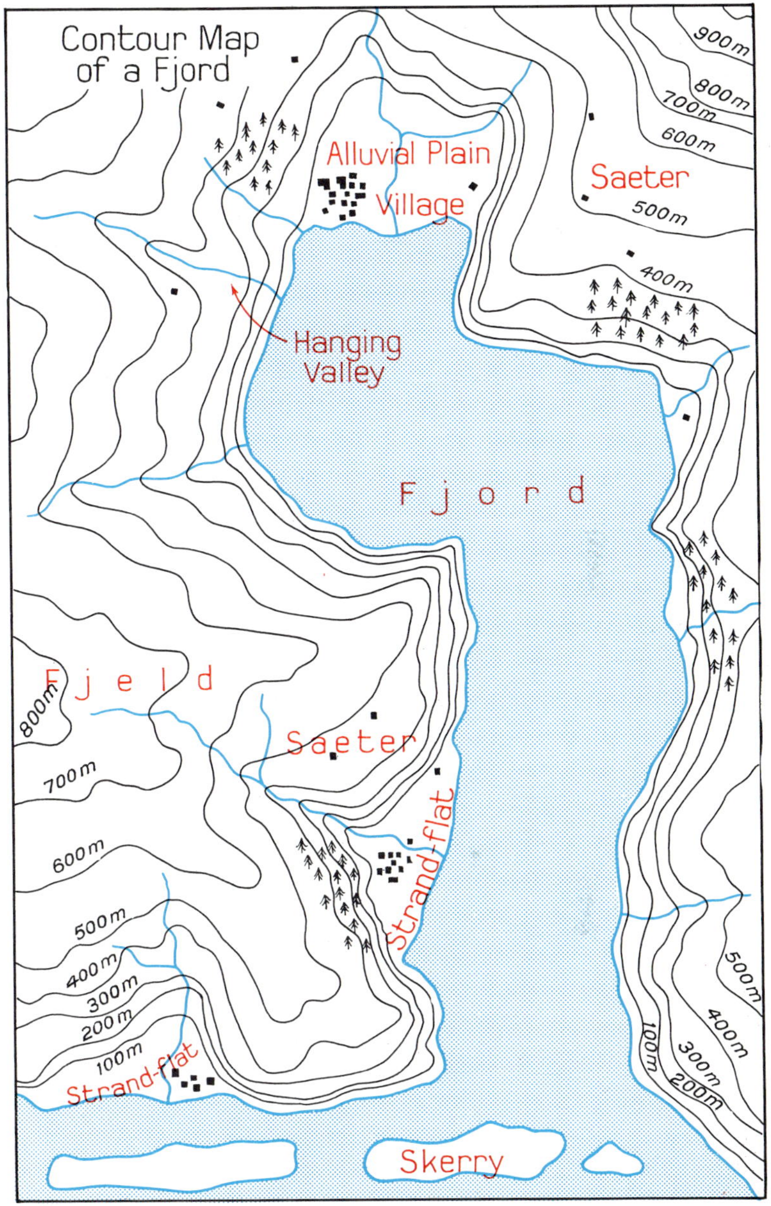

FIG. 3.5 **Contour map of a fjord**

23

## Northern Norway, from Trondheim to North Cape

This northern portion of the country is of varying width, narrowing to only 6 kilometres in the vicinity of Narvik, but widening to 160 kilometres in the extreme north. In the interior, the plateau is said to resemble the keel of an upturned boat, hence its name, the Kiolen Mountains. It is mostly uninhabited, save for some wandering Lapps who rely on their herds of reindeer for a living.

The coastline contains some impressive fiords and is fringed by a number of mountainous islands including the Lofoten group. The climate is mild and harbours are ice-free, in spite of the fact that the region extends well beyond the Arctic Circle. There are many small coastal settlements both on the mainland and islands where fishing is the chief occupation. The seas round the Lofoten Islands are exceptionally rich in cod, and the drying of cod and cod liver oil extraction are carried on in most of the towns and villages. *Tromso* and *Hammerfest* are bases for the Arctic sealing and whaling fleets, but most whaling is now carried on in the Antarctic from bases in Southern Norway. *Narvik* is particularly concerned with exporting Swedish iron ore.

In recent years new industries based on hydro-electric power have been developed in the region. Some low-grade iron ores are mined near Narvik and at *Kirkenes*, near the Russian frontier, and a large integrated iron and steel works has been built at *Mo-i-Rana*. Here a town of 20,000 inhabitants has already grown up in a short time, and eventually the works will provide 50 per cent of the steel that Norway needs. Nearby are other factories including aluminium works. The main means of communication along this coast is by steamer, but contact with the main towns is also maintained by regular air services.

**Spitzbergen:** This is a Norwegian island 800 kilometres north of North Cape. It is mostly covered by ice or has a tundra vegetation, and the subsoil is permanently frozen. There is some coal mining, and fishing from harbours in the west, which are open for six months of the year.

## Exercises

*Answer in note form:*

1. Explain the following terms – Fiord, Strandflat, Skerry, Saeter, Fjeld.
2. Why is northern Norway sometimes described as the 'Land of the Midnight Sun'?
3. What types of fish are caught off the Norwegian coast, and why are they found there?
4. Describe and locate the metallurgical industries of Norway.

*Essay Question*

How have the relief, climate and natural resources influenced the occupations of the Norwegians?

**Fertiliser plant on Glomfjord**　　　*Royal Norwegian Embassy*

Here nitrates, just inside the Arctic Circle, are produced using atmospheric nitrogen and locally produced hydro-electricity.

## Narvik

*J. Allan Cash*

Every day over 30 iron ore trains reach Narvik from the Swedish mines, mostly from Kiruna. The ore is taken to the stockyards from which it is moved to the ships on conveyor belts. The capacity of the ore harbour now exceeds 20 million tonnes a year.

## A fishing village on the northern coast   *J. Allan Cash*

Compare this scene with that of Narvik. Note the barren nature of the coast, the small fishing craft and the processing sheds lining the quays.

# 4: SWEDEN

Sweden is mountainous in the north and west where the Scandinavian Highlands form the frontier region with Norway. From here the land slopes gently to the south-east, coming up against the ancient crystalline rocks of the Baltic Shield, and ending in a coastal plain bordering the Gulf of Bothnia (Figure 4.1).

South of latitude 60°N the topography becomes more broken and there is a large down-faulted depression known as the Central Trough, containing lakes Vaner, Vatter and Malar. To the south the land rises again to the plateau of Smaland, and then falls away to a fertile plain to which the name Scania has been given.

Almost the whole of Sweden was under a thick ice sheet during the Ice Age, so that most of the lowlands have been left with a mantle of glacial deposits, chiefly of gravel and boulder clay.

Sweden's climate is more continental than that of Norway. The winters are more severe and the summers are slightly warmer in comparable latitudes, whilst the rainfall is generally less than 750 mm per annum. Of course conditions vary from north to south; in the extreme north, in Lapland, the winters are long and the growing season lasts only about three months, whilst in Scania in the south it lasts seven months.

The area of Sweden is about 450,000 square kilometres and the population is just over 8·2 million, giving a density of population of some 18 to the square kilometre, which is 50% more than that of Norway. As in Norway, the distribution of population is very uneven, and most people live in the Central Trough, particularly in and around the two great cities of Stockholm and Gothenburg, or along the south coasts.

## Agriculture

Just over 8 per cent of the land is farmed, and much of this is in Scania and the Central Trough. The farms are small but highly mechanised, and use large quantities of fertiliser. The number of farm workers has declined rapidly in recent years, but production has remained steady as farm techniques have improved. Stock farming is normally the main activity, especially dairying, though there

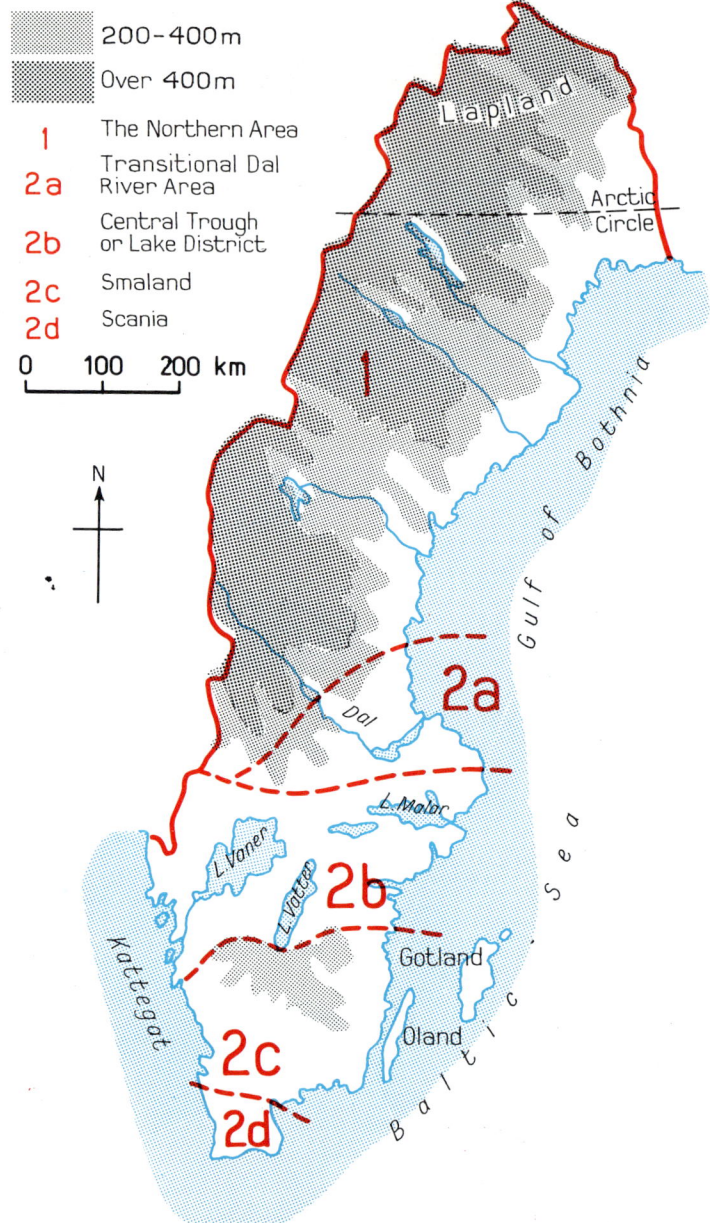

200-400m
Over 400m

1    The Northern Area
2a   Transitional Dal River Area
2b   Central Trough or Lake District
2c   Smaland
2d   Scania

0   100   200 km

FIG. 4.1 **The regions of Sweden**

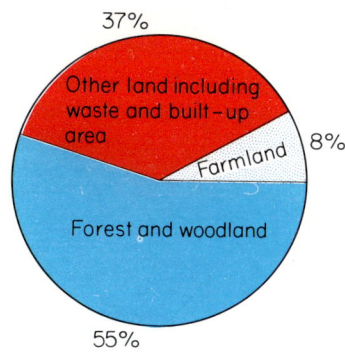

FIG. 4.2 **Land use of Sweden**

are also many beef cattle and pigs. Arable crops include wheat and sugar beet in Scania, oats, barley, rye and fodder crops in central Sweden, and potatoes almost everywhere. In all, Sweden produces about 90 per cent of its own food.

Forests cover 55 per cent of the land surface. From just north of the Central Trough right up to the Arctic Circle is a great belt of coniferous forest, consisting mainly of pine and spruce trees, whilst in the south the wooded areas are more broken, with both coniferous and deciduous trees, the latter including beech and oak. Lumbering has for long been a major industry, helped by the numerous rivers which provide a means of transport to the sawmills and pulp mills. Much of the timber is used in the manufacture of pulp and paper, rayon, matches, doors, window frames, furniture and plywood, and considerable quantities of timber and timber products are exported.

### Mining and Industry

Besides timber, the other major natural resource is iron ore. There are rich deposits of high grade ore in the far north at Kiruna and Gallivare, and in the central area in the Bergslagen. Many new ore bodies have recently been discovered including copper, lead, zinc, nickel and iron in the vicinity of Skelleftea. In recent years over 85 per cent of the country's iron ore production, which has totalled nearly 30 million tonnes per annum, has been exported.

Sweden lacks both coal and mineral oil, and her power supplies are derived mainly from numerous hydro-electric stations. Conditions for generating hydro-electricity are less favourable than in Norway, since the more subdued relief restricts the fall of the rivers and many of them freeze in winter. Eighty per cent of the water power resources are

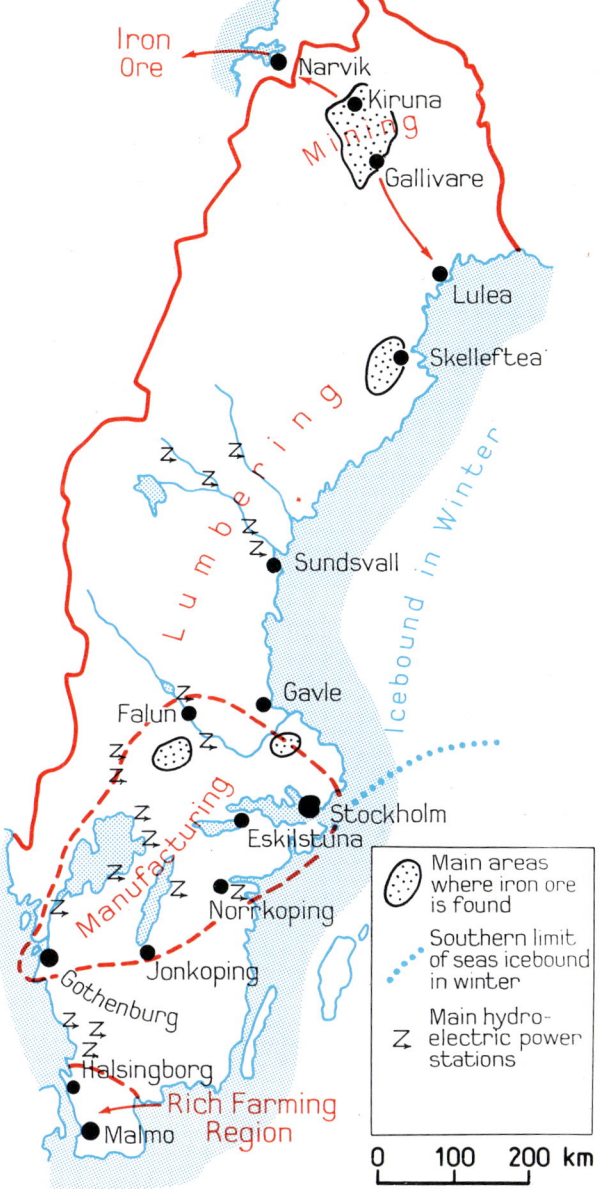

FIG. 4.3 **Sweden—economic**

in the north, but the greatest demand for power is in the centre and south, so that large quantities of electricity must be conveyed southwards by high voltage grid. Submarine cables carry some of the electricity across the Sound to Denmark. The Swedes have made use of the exploitable rivers to such an extent for hydro-electricity that, under present conditions, there are very few potential sites left.

The strength of Sweden's economy lies in its iron and steel, and engineering industries, which are widely spread through the centre and south of the country, with the main concentrations at Eskilstuna and in the Stockholm area. There is an emphasis on the production of high quality steels for use in the engineering industry, which includes precision and electrical engineering, shipbuilding and the car industry. In shipbuilding, Sweden's production in recent years has been greater than that of Britain or West Germany, and includes a large number of tankers, mainly for Norwegian customers. The growth of the motor car industry is even more remarkable, expansion being most marked in the sixties.

The chemical industry, making use of a wide range of raw materials including timber and petroleum, has also made rapid progress, along with the textile industry, which is carried on in coastal towns since it is based mainly on imported raw materials, and the manufacture of prepared foods, of which Findus products and Ryvita are known throughout western Europe.

In spite of the lack of coal and mineral oil, Sweden has become an advanced industrial nation, with one of the highest standards of living in the world. This remarkable achievement can be attributed to (1) the abundance of iron ore, timber and hydro-electric power, (2) the fact that the country is almost self-supporting in food, (3) the efficiency and inventiveness of Swedish workers (their inventions include safety matches, cream separators and lighthouse lamps), and (4) a policy of neutrality, whereby it has avoided destructive wars.

## THE REGIONS OF SWEDEN

### Northern Sweden

Figure 4.4 shows the main physical features of Northern Sweden. Much of the surface is covered with coarse morainic material and there are many parallel rivers flowing south-east, with long lakes in their upper courses, some held up behind moraines and others occupying rock basins carved by glaciers. The only fertile area is the coastal plain which is largely covered by alluvial sediments. A few hardy crops are grown here and cattle are kept, despite the long, severe winters which

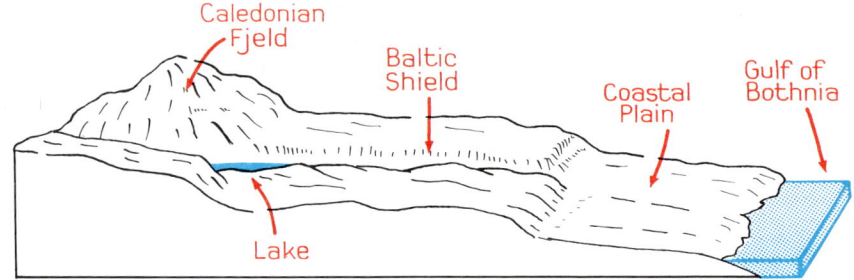

FIG. 4.4 **A block diagram of a portion of North Sweden**

cause the Gulf of Bothnia to freeze for five months. There is tundra on the fjeld and in the far north, but much of the rest of the land is covered with pine and spruce forest, and lumbering is the main activity. The logs are floated down the rivers to sawmills and pulp mills, and the many waterfalls provide hydro-electric power for these and other industries based on the timber. One of the main timber working towns is *Sundsvall*.

The iron ore deposits of *Kiruna* and *Gallivare* are among the richest in the world and there are very large reserves. The ore is found in hills which have resisted erosion, and is a magnetite, with a metal content of between 60 and 70 per cent. The miners are given high wages and good living conditions, as a recompense for the long winters with very short periods of daylight. *Lulea*, which has an iron and steel plant, is the chief port on the Gulf of Bothnia for shipping the ore, but most is also exported via Narvik in Norway, which is ice-free throughout the year. South of Lulea is *Skelleftea,* a new centre for mineral working where iron and non-ferrous ores, including copper, lead, zinc and nickel, occur. Altogether 80 per cent of Sweden's iron ore comes from these northern areas.

![Kiruna mining town photograph]

**Kiruna**

*LKAB Iron Ore Company*

The arctic mining town of Kiruna is overlooked by the Kirunavaara iron ore mountain where open-cast mining has now been replaced by underground working. The photograph was taken during one of the long days of summer.

**Mining iron ore**   *LKAB Iron Ore Company*

Mechanisation is the key-note of modern mining, and the miners are equipped with more and more complex and automated machinery, exemplified in this drift drilling rig which is used in the Kirunavaara underground mine.

**Reindeer round-up in Lapland**   *Swedish National Travel Association*

Reindeer are kept in northern Sweden, and the photograph shows a round-up, when the different herds are sorted out and a certain number selected for slaughter. The herdsmen are mainly Lapps.

*Svenska Cellulosa Aktiebolaget, Sundsvall*
Study these photographs which show a power saw at work, machinery for stripping branches, a pulp mill and newsprint machines. Then describe the various stages in the production of pulp and paper. Include transport facilities and power supplies.

## The Dal River Area

The break from the forested, bleak north to the throbbing life of the Central Trough is gradual. The Dal river area represents this transitional zone. Here mining, agriculture, manufacturing and forestry are all in evidence. Rich haematite iron ore is mined in the Bergslagen district and is used locally in the production of high quality steel. *Falun* and *Dannemora* are two of the main centres of steel manufacture. Formerly copper was mined round Falun, but it is now almost worked out, though there is some production of pyrites for the chemical industry.

## The Central Trough

This is the heart of Sweden, with Stockholm as the eastern focus and Gothenburg the western. It is a faulted area with basins filled with clays or glacial gravels. Blocks of ancient crystalline rocks have been uplifted to form low hills, and these as well as much of the gravel lowland are covered with conifers. After the Ice Age the region was submerged by the sea, and then there followed a period of uplift leaving the lakes occupying the hollows. Agriculture is limited mainly to the clay soils, with an emphasis on dairying and the growing of oats, barley, rye and hay, chiefly for fodder.

Manufacturing is dominant throughout the region. It developed from domestic industries using water power. During the nineteenth century there was some movement to the coast when imported coal came to be used, because land transport for such a bulky commodity was expensive. However industry has become more widely dispersed during this century as electricity obtained from hydro-electric as well as thermal sources was increasingly used. Engineering, textile, chemical and woodworking industries are well represented. *Norrkoping* is the main textile centre, producing goods made from both synthetic and natural fibres. *Jonkoping* is a leading engineering town and also the centre of the Swedish match industry. *Eskilstuna* is noted for high quality steel production including cutlery.

*Stockholm* (800,000) is the country's capital and main commercial and industrial centre. It began as a fortress on an island in the narrow entrance to Lake Malar, and later spread to both shores. It lies 50 kilometres from the Baltic, at an important bridging point and route

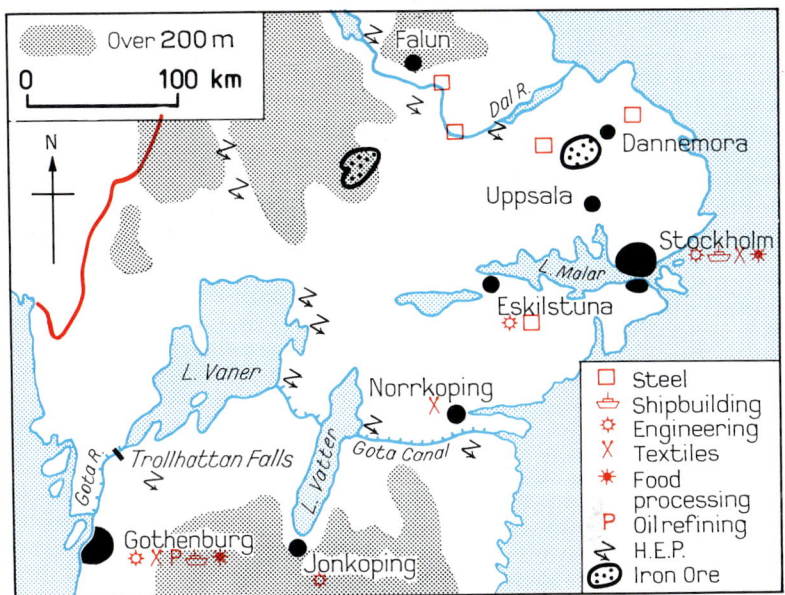

FIG. 4.5 **The Central Trough**

focus. It has developed a wide range of industries including textiles, engineering, shipbuilding and consumer goods, and is Sweden's second port, although icebreakers are needed to keep the harbour open in winter.

*Gothenburg* (450,000) lies on the Kattegat, and is the largest port, being more accessible from the North Sea than Stockholm. It is ice free and the mouth of the Gota river provides a deep, sheltered harbour. It has many industries including engineering, oil refining, textiles, glassware, food processing and shipbuilding. The latter has become of great importance in recent years and relies on highly automated processes. The hydro-electric station at the Trollhattan Falls provides the city with much of its power. The Gota Canal provides a link with Stockholm via Lake Vaner and Vatter, but is now used mainly by pleasure craft.

31

**Stockholm**

*Swedish National Travel Association*

Stockholm is often described as the Venice of the North. The oldest part of the city lies on an island shown in the centre of the photograph, whilst the new residential and industrial areas have developed on the shores of Lake Malar.

## Smaland

South of the Central Trough is the low plateau of Smaland which is made of hard crystalline rocks including granite. It is mainly heath and moor, with stretches of woodland, and agriculture is only possible on the marine clays of the coastlands. The well-known Swedish glassware is produced on the fringe of the upland, using local pure sand.

Two islands, Oland and Gotland, lie off the east coast. Gotland is the more prosperous, with cattle and market gardening, whilst Oland has some farming and a number of fishing centres.

## Scania

Although it occupies only a small portion of Sweden, this densely-populated lowland plays an important part in the country's economy. Because of its mild climate and fertile boulder clay soils, it is the only part of the country where wheat and sugar beet grow well, and it also has the greatest concentration of dairy cattle and pigs. *Malmo* (250,000) is Sweden's third city, and has easy access across the Sound to Copenhagen. It has an artificial harbour and has developed a similar range of industries to those of Gothenburg, including the processing of imported raw materials and foodstuffs, engineering, shipbuilding and textiles.

**A panorama in North Sweden**          *Refot, Stockholm*

**Village in Scania**          *Refot, Stockholm*

What are the main differences between the two scenes? What are the reasons for these differences?

## Exercises

*Answer in note form:*

1. Compare the distribution of forests in Sweden with that in Norway.
2. Why is Narvik ice free in winter, whilst Stockholm, which is much further south, needs icebreakers to keep its harbour open?
3. Compare and contrast the sites, positions and functions of Stockholm and Gothenburg.

*Essay Question*

Write an explanatory account of (a) metal working, and (b) forestry, in Sweden.

# 5: DENMARK

Denmark consists of the peninsula of Jutland, several large islands and some hundreds of small ones. Physically it is part of the North European Plain, and the underlying chalk is hidden almost everywhere by thick layers of glacial material. Quite different in character is the small island of Bornholm in the Baltic, which is composed of ancient granite rocks.

Crossing Jutland from north to south is a great terminal moraine left behind after the retreat of the Baltic ice sheet, forming hills nearly 200 metres above sea level. To the west the land is covered with sandy outwash material which was carried away from the ice sheet by melt water, and the coast is bordered by sand dunes with many shallow lagoons. Eastern Jutland, on the other hand, is covered with boulder clay of the ground moraine and has a number of winding coastal inlets locally called fiords, although they have little in common with the Norwegian fiords. North Jutland is cut across by Lim Fiord and ends in a sandy peninsula called the Skaw.

The main islands, which include Funen, Laaland and Zealand, lie across the shallow entrance to the Baltic Sea, and are the remains of a plain which stretched from Jutland to southern Sweden. The plain was partially drowned as a result of the rise in sea level after the Ice Age, when the Little Belt, Great Belt and The Sound were formed. The islands have a gently undulating surface and are mainly covered with boulder clay.

The climate of Denmark is similar to that of eastern England, but with slightly colder winters showing the greater continental influence. The rainfall is light, ranging from 500 to 700 mm per year. Copenhagen has an annual rainfall of 560 mm, and mean temperatures of 18°C (July) and 0°C (January).

With a population just over 5 million, and an area of 43,000 square km, Denmark is by far the smallest and most densely populated of the Scandinavian countries. It is not well endowed with natural resources, for it has no coal or mineral ores, no water power, and its soils are not naturally very fertile. Yet by the efforts of its people it has developed a prosperous economy and the country has a high standard of living.

## Agriculture

Denmark's achievements in agriculture are outstanding. In the nineteenth and first half of the twentieth centuries the country's prosperity was based almost entirely on farm production. Until the 1860's a system of mixed farming was in general use, and one of the main exports was wheat. Then came the competition of cheaper wheat from the Canadian Prairies and Russian Steppe. As a result, the Danish government persuaded its farmers to concentrate on livestock.

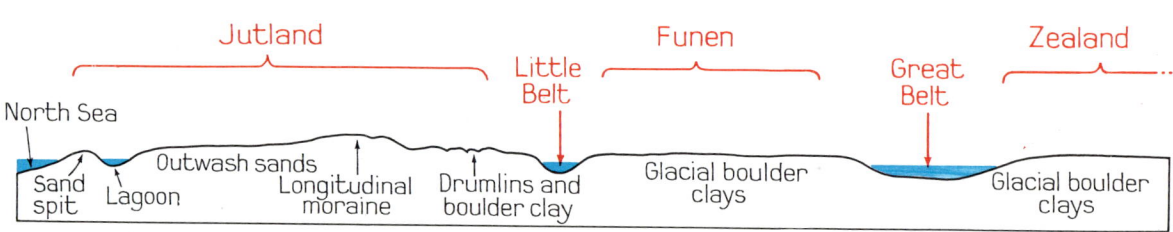

FIG. 5.1 **Section across Denmark**

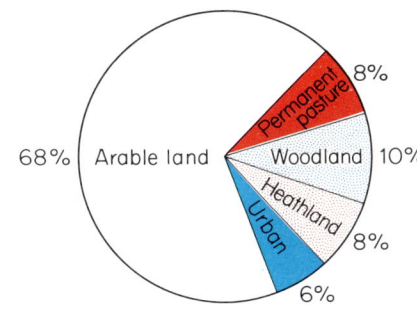

FIG. 5.3 **Land use of Denmark**

Today about 76 per cent of the land is farmed, which is a very large proportion compared with most European countries. Almost everywhere there are small, family farms, and intensive, scientific methods are employed giving yields per acre which are among the highest in the world. The main basis of farming is a small dairy herd together with pigs and poultry. Most of the land is devoted to growing fodder crops, particularly barley, oats and roots, with only a small proportion under grass. The cattle are seldom allowed to graze, but are fed in the

**A farming scene**       *Royal Danish Embassy*

Note the typical small, white farmhouse and the British made tractor.

FIG. 5.2 **Denmark**

stall or farmyard. Under this system far more animals can be kept per hectare than if the land were under grass. To supplement the home-produced feeding stuffs, large quantities of maize and cattle cake are imported. Many farmers also set aside some land for cash crops such as wheat, sugar beet and potatoes, and some specialise in market gardening.

The co-operative movement is well-developed in Denmark, and deals with the processing and marketing of farm produce, as well as the supply of machinery and fertilisers. There are co-operative creameries, bacon factories and egg-packing stations, to which the farmers send their milk, pigs and eggs. Quality is maintained by regular government inspection, and government-run educational and advisory services are available for the farmers.

Regional differences in farming are not very marked, although in general it may be said that the islands and eastern Jutland, with their fertile boulder-clay soils, provide the most valuable farmland. The least fertile areas are in western and northern Jutland, but even here much of the land has been reclaimed and improved. A century ago the sands and gravels were mostly covered by heath and moorland, and there were peat bogs in the depressions. Reclamation has involved the marling of the lighter, sandy soils i.e. the mixing of the sands with marl (a compacted, fine silt) to improve the soil texture and moisture-holding capacity. Many of the inland bogs and coastal marshes have been drained and the peaty soils heavily limed. Bornholm, distinctively, has much moorland on the granite, especially in the north where sheep grazing is the main farming activity. In the south is some arable farming with oats as the chief crop.

In recent years the expansion of manufacturing industries in Denmark, and a decline in the number of agricultural workers has meant a marked change in the balance of the economy. Today three quarters of the population are town-dwellers, and only 9 per cent of the working population are directly engaged in farming. Nevertheless agriculture still provides a considerable proportion of the country's exports, the chief items being bacon, pork, butter, cheese and eggs. The traditional markets for these products are still Britain and, to a lesser extent, West Germany.

◄ **Railway viaduct over the Little Belt**    *Royal Danish Embassy*
Note the gently undulating landscape and the scattered small farms on the island of Funen.

## Fishing

The fishing industry has grown rapidly during this century. Along the shallow west coast of Jutland, plaice and sole are caught, whilst cod, herring and mackerel are fished in the Kattegat and North Sea. Herring is the biggest catch, but plaice the most valuable. The Lim Fiord in recent years has been used as a hatchery and breeding ground, especially for plaice and oysters. Esbjerg and Skagen are the chief fishing ports.

## Manufacturing and Towns

Among the factors favouring the development of industry are Denmark's position on the sea routes between the North Sea and Baltic Sea, an efficient system of internal communications with train ferries or long bridges connecting the main islands, and an adaptable, well-educated labour force. For power supplies the country depends on imported coal, mineral oil, and electricity (by submarine cable from Sweden), and many of the raw materials also have to be imported.

The most important branch of manufacturing is metal working, which employs one quarter of all the industrial workers and provides the largest amount of exports. There are steelworks at Frederiksvaerk, using scrap iron and Swedish iron ore. Large quantities of steel are used in shipbuilding, mainly at Copenhagen and Elsinore, and in various branches of the engineering industry including marine engineering, electrical engineering, and the making of agricultural and dairy equipment.

Next in importance both for production and export is the processing of foodstuffs, particularly bacon curing, sugar refining, and the manufacture of butter, cheese and condensed milk. Imported oilseeds are crushed for the making of cattle cake and margarine. Other leading industries are the manufacture of cement, using the local chalk in north Jutland, chemicals, textiles, furniture and porcelain.

Most of the towns are small and function as market centres with food processing industries, or have developed as ports along the coast. *Aarhus* (245,000) is the second city of Denmark and has a good natural harbour. It is a commercial centre serving the surrounding farming land, and has engineering, shipbuilding and clothing industries as well as food processing. *Aalborg* is another port, lying on Lim Fiord, and is the main centre of the cement industry. Its other industries include the manufacture of fertilisers and the crushing of

oilseeds. *Esbjerg,* which has an artificial harbour, is the leading port of the west coast. It is important particularly for its trade with Britain and as the country's leading fishing port. It has a direct railway link with Copenhagen, via the train ferry over the Great Belt.

**Esbjerg—the fishing port** *Royal Danish Embassy*

### Copenhagen

Locate the ports of Copenhagen and Esbjerg on a small sketch map and then give as many reasons as possible for the differences between them.

*Copenhagen* (1,250,000) is the capital of Denmark and the largest city of Scandinavia. It stands on the east coast of Zealand and dominates the Sound, the deepest and most used entrance to the Baltic. In the past its position was of strategic importance, and a toll was charged for ships passing through the Sound. Today it handles most of Denmark's trade and is an entrepôt port serving the Baltic countries. There is a ferry service to Malmo in Sweden.

The city's expansion during the twentieth century resulted from the growth of manufacturing, and forty per cent of the nation's industrial production is now centred in Copenhagen. Shipbuilding and marine engineering, electrical and other branches of light engineering, the making of agricultural implements and food processing are among its main industries. Royal Copenhagen China, for which kaolin is brought in from the island of Bornholm, and Carlsberg and Tuborg lagers are particularly well known products.

About 30 kilometres to the north of Copenhagen is the town of *Elsinore*, whose castle was the setting for Shakespeare's 'Hamlet'. It is mainly a holiday resort, but also has shipbuilding industries, and there is a ferry service from Elsinore to Halsingborg in Sweden.

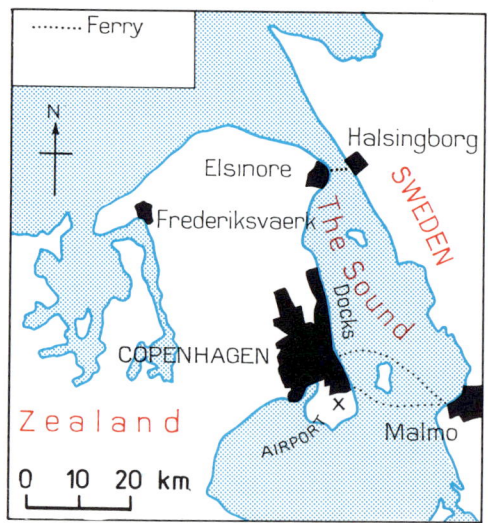

FIG. 5.4 **The position of Copenhagen**

# Exercises

*Answer in note form:*

1. What physical factors account for the differences in scenery between west Jutland, Zealand and Bornholm?
2. Why does Copenhagen have a large entrepôt trade?
3. Describe a typical Danish farm on Zealand.

*Essay Question*

How have geographical factors influenced the type of farming in Denmark?

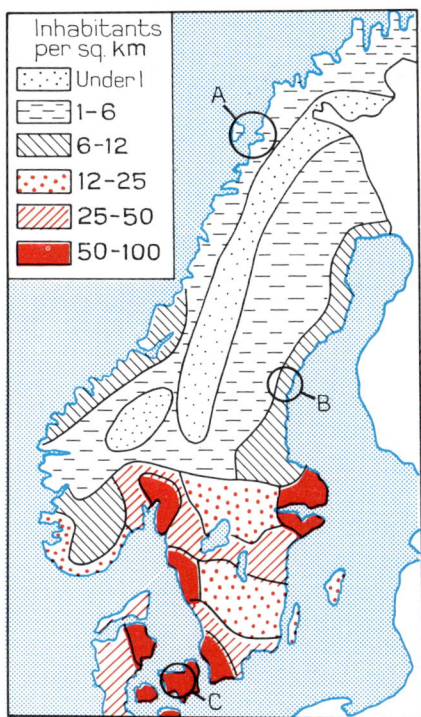

FIG. 5.5 **Distribution of population in Scandinavia**

How have the local geographical conditions influenced the population densities at A, B and C?

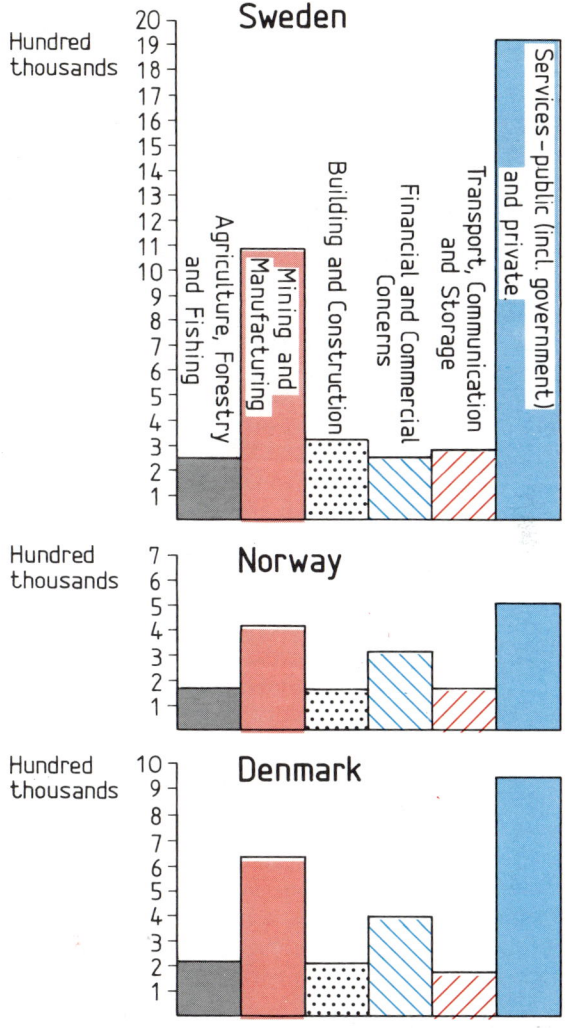

FIG. 5.6 **Main sectors of employment in Scandinavia**

1. Which of these sectors (a) are concerned with primary production, (b) are concerned with secondary production, and (c) are tertiary activities?
2. Estimate from the graphs the total employed in each sector for each country.
3. Point out the significant differences between the sector totals of the three countries.

# 6: THE NETHERLANDS

The Dutch refer to their country as the Netherlands, which means 'low-lying lands', or alternatively as Holland (the 'hollow land'), although strictly speaking the latter refers only to the region between the Rhine and the peninsula of Den Helder.

The Netherlands lie where three great rivers, the Rhine, Maas and Scheldt reach the sea and have formed a vast delta. In the Middle Ages marshland must have extended in a wide belt along the coast and up the broad river valleys, whilst in the east there was a waste of gravels and infertile sands left by ice sheets after the Ice Age. Man lived on mounds rising from the marsh or on higher land to the east. As the marshes became filled with silt and reeds, dykes were built to keep out the high tide, and drainage ditches and canals were dug to carry away the surplus water. Nature had already provided a barrier of sand dunes along the coast and these were made the basis of the dyke system. As early as the seventeenth century large areas had been drained with the aid of windpumps, but progress was slow until the introduction of steam pumps in the nineteenth century, when the large area in Holland known as Haarlem Mere was reclaimed. Later came the development of electrical and diesel pumps so that even more ambitious schemes could be undertaken, and these included the reclamation of the Zuider Zee which was begun in 1927 and is now approaching completion. Today about half of the Netherlands consists of reclaimed land, much of it below sea level. The drained areas, which are known as polders, are the most fertile and densely populated parts of the country.

Great efforts are also being made to reclaim the large expanses of peat bog and heather moor in the eastern Netherlands. Many of the peat bogs have been drained by the digging of canals, and over large areas the peat has been mixed with the sand, making it possible to grow grasses and nitrogenous plants. In this way, and with heavy applications of fertilisers, the soil has been made quite productive.

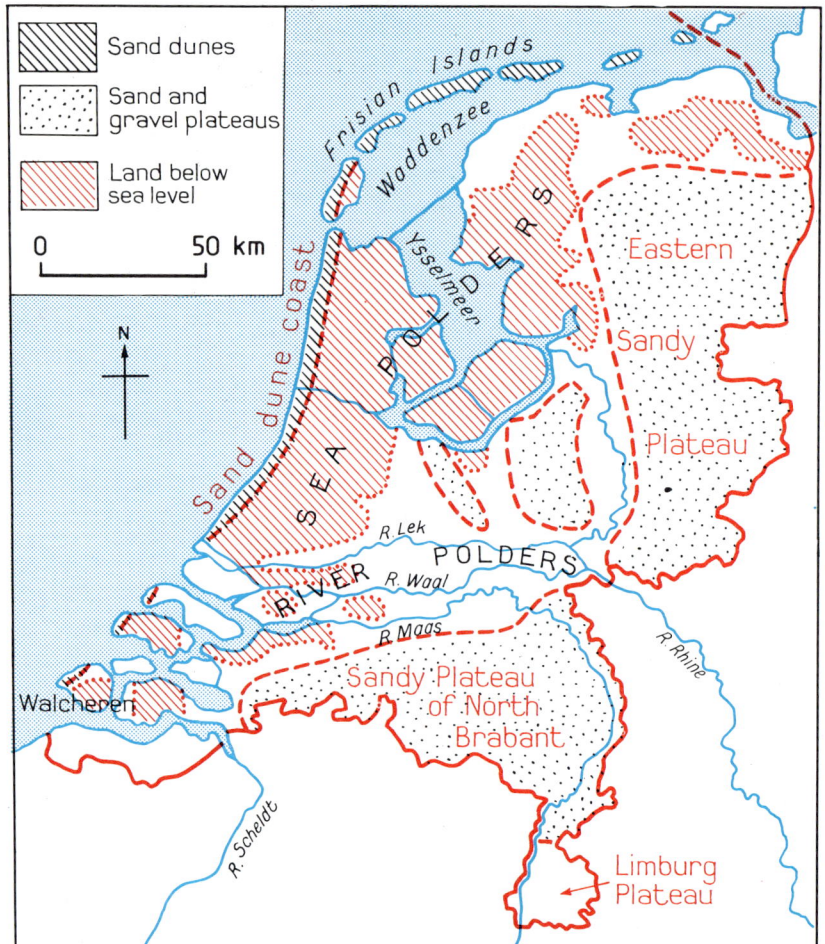

FIG. 6.1 **The regions of the Netherlands**

The climate of the Netherlands is typically West European in type, being slightly more extreme than that of eastern England, with mean January temperatures just above freezing point and mean July temperatures of about 17°C. The rainfall is fairly evenly distributed throughout the year, with amounts varying from 550 to 750 millimetres per annum.

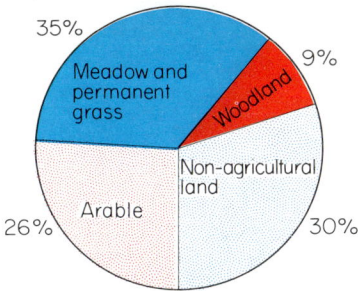

FIG. 6.2 **Netherlands—land use**

The Netherlands has a population of 13·7 million and an area of only 41,200 square kilometres, and is the most densely populated country in the world, with an average of about 330 people to the square km. Before the Second World War it was a great colonial power with world-wide commercial interests, and even today, after the loss of nearly all its colonies, it remains a major maritime and trading nation. It benefits greatly from its nearness to the busy North Sea shipping lanes, and above all from its position at the mouth of the Rhine, with the result that it handles much of West Germany's overseas trade. Its own industry and trade have also been helped by the economic union between Belgium, the Netherlands and Luxembourg, known as Benelux, and by membership of the European Economic Community.

## Agriculture

Agriculture employs about six per cent of the working population and about 60 per cent of the land surface is farmed. In recent years there has been a marked decrease in the number of agricultural workers, but farm production has been maintained by the greater use of machinery, fertilisers and scientific aids. The drift from the land has led to an increase in the size of holdings and this too has led to greater efficiency.

Over half the farmland, including the damper polders, is permanent pasture with dairying as the dominant activity. Arable farming is more important on the better-drained land and lighter soils, and the chief crops are wheat, sugar beet and potatoes, with market gardening especially important along the western edge of the polders. Dutch farmers have a reputation for heavy yields and high quality produce, and are able to provide most of the food needed by the country, as well as considerable exports of dairy produce, vegetables, fruit, flowers and bulbs.

41

**Polderland along the Drecht** ►

*Dutch Embassy*

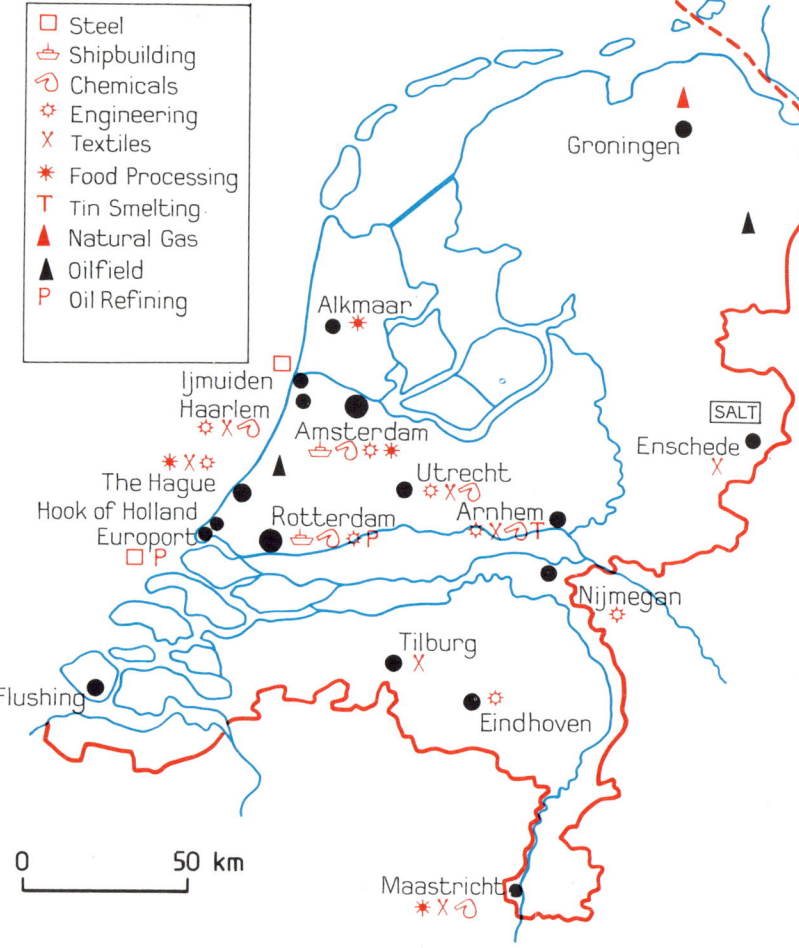

|  |  |
|--------|---------|
| □ | Steel |
| ⚓ | Shipbuilding |
| ⌔ | Chemicals |
| ✿ | Engineering |
| X | Textiles |
| ✳ | Food Processing |
| T | Tin Smelting |
| ▲ (red) | Natural Gas |
| ▲ (black) | Oilfield |
| P | Oil Refining |

FIG. 6.3 **Industry in the Netherlands**

## Fishing

Fishing is a traditional occupation, although it is no longer very important. Most of it is done from small craft in home waters or in the North Sea, and herrings make up half the catch. Shellfish, including mussels, are found in shallow waters near the coast. The chief trawler port is Ijmuiden.

## Industry

Industry has made rapid progress in recent years. Before the last war it was based mainly on local agricultural produce and imported raw materials such as oil seeds, cotton, timber and rubber, chiefly from the Dutch colonies. Other domestic resources are now proving of considerable value. There are newly-developed oilfields in Drenthe and south Holland which satisfy about one third of the home demand. More important still was the discovery, after the war, of a huge natural gas field in the Groningen area. Today this field makes a massive contribution to the Dutch energy requirements and a significant surplus is exported to neighbouring countries.

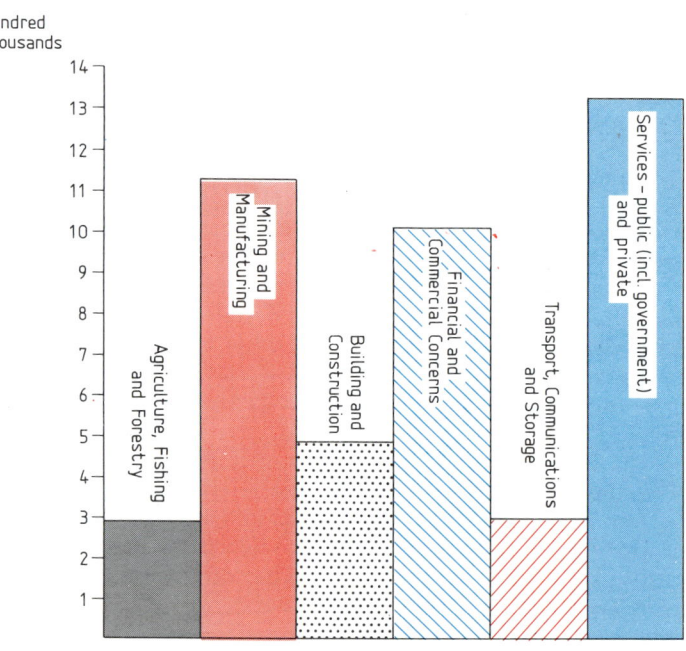

FIG. 6.4 **Main sectors of employment in the Netherlands**

**Ijmuiden steelworks**

*Dutch Embassy*

The ease of importing raw materials, and the excellent network of communications by road, rail and inland waterway, have greatly helped industrial development. Other favourable factors are the abundance of skilled labour and the large populations of neighbouring countries which provide a ready market for manufactured goods.

The metal and engineering industries are the most important and most rapidly expanding. Iron and steel production is concentrated at Ijmuiden and at new works at Europort, both well-placed for the import of iron ore. There is a large tin smelting plant at Arnhem, and a new aluminium works in the far north, using local supplies of natural gas. Shipbuilding, ship repairing and marine engineering are carried on at the main ports, whilst other important branches of engineering include the manufacture of electrical equipment, motor vehicles and land drainage equipment. There has also been expansion in the textile and chemical industries, as well as in food processing, which includes butter and cheese-making, the canning and quick-freezing of fruit and vegetables, and the manufacture of margarine and chocolate.

The Netherlands is a country with a great cultural heritage, particularly with respect to painting, and the art gallery at Amsterdam, with its collection of Rembrandt and Vermeer paintings, is an outstanding tourist attraction. In spring many tourists also come to view the bulb fields which cover thousands of acres in Holland.

FIG. 6.5 **The provinces of the Netherlands**

## THE REGIONS OF THE NETHERLANDS

### The Polderlands

The polders are most extensive in North and South Holland, where the coast is fringed by sand dunes which have been strengthened to protect the low-lying land from the sea. Farther north, the line of dunes is continued by the sandy Frisian Islands, the islands being separated by shallow water from the coastal polders of Friesland and Groningen. To the south of Holland lies the delta region of Zeeland, consisting of many islands whose polders have to be protected from flooding by both the sea and the rivers. Inland, polders follow the maze of watercourses including the Lek, Waal and Maas, which are dyked along almost their whole lengths.

Mention has already been made of the reclamation of the Zuider Zee. A 29 kilometre-long sea wall across the entrance was completed by 1932, and five polders have been created, leaving a large fresh-water lake called the Ysselmeer. Much of the newly-won land is already being farmed and the scheme, when completed in 1980, will have increased the cultivated area of the Netherlands by 10 per cent. Shipping can still reach Amsterdam, using lock gates in the sea wall.

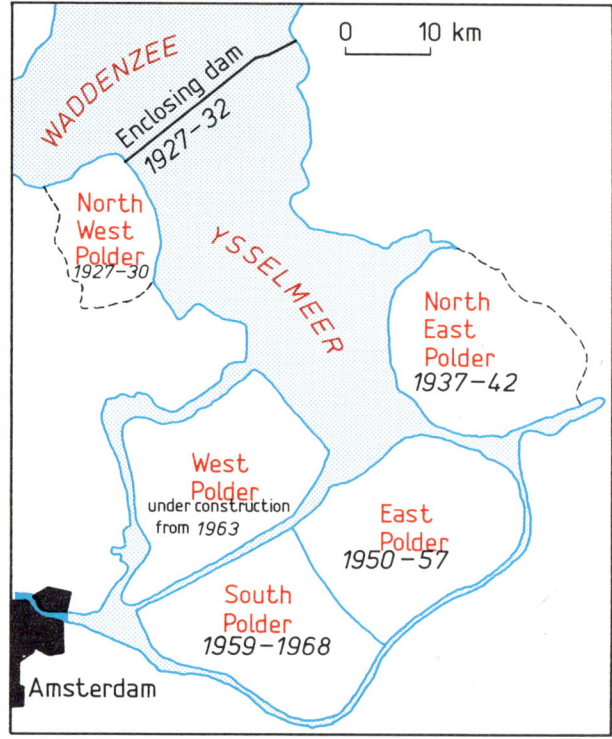

FIG. 6.6 **Stages in the drainage of the Zuider Zee**

**Wieringermeer Polder with Ysselmeer** *Dutch Embassy* ▶
Describe the methods of land reclamation illustrated in this photograph and the one of polderland along the Drecht and state how the land is now being used. Illustrate your answer with simple plans based on the photographs.

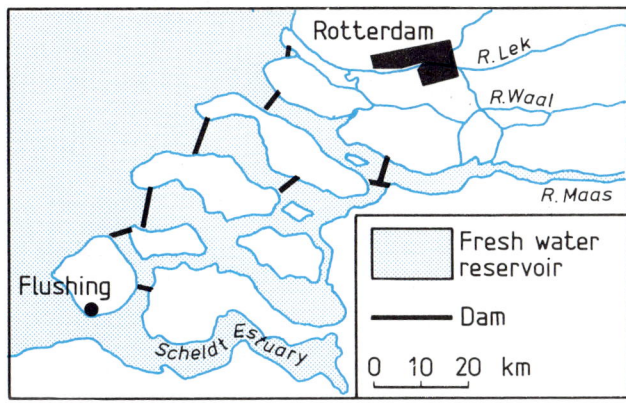

FIG. 6.7 **The Delta Scheme**

There have been many disasters in the past when the sea has broken through the dykes and invaded the land. The latest of these was in 1953 when very high tides, combined with gale force winds, caused widespread flooding especially in Zeeland, with the loss of 1,800 lives. In order to make sure that this cannot happen again, the 'Delta Plan' was drawn up, by which the sea will be kept out of the delta region by the building of dykes across four tidal estuaries (Figure 6.7). The dykes carry roadways and help to stabilise the shifting sand: the large fresh water lakes which accumulate behind them provide valuable additional supplies of fresh water. The work was begun in 1956 and is nearly complete.

The damper polders, especially in Holland, in the lower parts of Zeeland, and along the rivers, are used mainly for stock farming, and dairy cattle, poultry and pigs are found on most of the farms. Large quantities of milk are sent to the towns of the Netherlands and nearby parts of West Germany. Much milk is also used in the making of butter, cheese and condensed milk, including the famous 'Gouda' and 'Edam' cheeses which are named after small market towns in Holland.

Arable farming is particularly important on the better-drained peaty polders, on the newly-reclaimed areas of the Zuider Zee, and on the heavy clays of the delta region, the chief crops being wheat, sugar beet and potatoes. Market gardening has expanded enormously during the present century in response to the increased demand for fresh vegetables from the densely populated industrial countries of North-West Europe. The chief market garden concentrations are in Holland along the old, worn down sand dunes and where the sand has been blown inland over the peat. The mixture of sand and peat provides an ideal soil for vegetables, fruit and flowers. Bulb growing is important round Haarlem, whilst south of the Hague much of the land is under glasshouses, producing tomatoes, lettuces, cucumbers and other crops.

## The Eastern Heathlands

To the east and south-east of the Zuider Zee, and to the south of the great curve of the Maas in North Brabant, are large expanses of glacial sand and gravel, forming undulating heathland between 50 and 100 metres above sea level. The natural vegetation consists mainly of heather, gorse, silver birch trees and poor pasture, with considerable stretches of peat bog in areas where the drainage is impeded.

For centuries the heathlands have provided little more than some rough grazing for sheep, but in recent years much work has been done to improve the sandy soils by mixing the sand with peat and applying fertilisers, and to drain the bogland. In this way large areas of land now produce good crops of potatoes, oats, rye and roots, and support cattle, pigs and poultry. Some of the least fertile land has been planted with coniferous trees which help to bind the soil, but there are still some stretches of bare sand. On the whole, the heathlands are by far the most thinly populated parts of the Netherlands.

## The Limburg Plateau

This region is in marked contrast to the rest of the country, and is a dissected plateau, built mainly of chalk, which rises to over 300 metres, although the average height is between 100 and 200 metres. Much of it is covered by loess which gives fertile, loamy soils. Good crops of sugar beet, wheat and potatoes are grown, and fruit farming, market gardening and dairying are other important branches of farming.

The plateau is underlain by Coal Measures at a depth of some 60 metres in the east but sinking to 600 metres in the west. Coal production has now ceased, but in the past, it helped the development of varied industries. Metal-working, glass, pottery and soap are important round the regional centre of Maastricht.

# THE TOWNS OF THE NETHERLANDS

If you refer to Figure 6.3 you will notice that the large centres of population extending from Rotterdam, The Hague, Haarlem and Amsterdam round to Utrecht form a great crescent of urban development which is known as the Randstadt. One of the most pressing problems of the Dutch Government has been the excessive concentration of people in this part of the Western Netherlands. In the Randstadt and surrounding areas live over 6 million people out of a total Netherlands population of 13·7 million. By contrast, the eastern and northern provinces are relatively empty and government policy is to encourage industrial expansion and new urban growth in Groningen, Gelderland, Friesland, Overijssel and Drenthe.

*Amsterdam* (1,036,000) grew up as a fishing port on a small creek leading into the Zuider Zee. Today it is the country's joint capital (with the Hague) and second port. The 24 kilometre-long North Sea Canal, which was completed in 1876, can accommodate the largest ships and provides a vital link with the coast at Ijmuiden. There are also canal links to the Rhine, to the north, and to the Ysselmeer. Amsterdam is a great financial and commercial centre, and has a wide range of industries including many which developed because of former colonial trade, for example the manufacture of rubber, tobacco, chocolate and margarine. It is the world's leading diamond market and has diamond-cutting and polishing industries reflecting the early links with South Africa. However the most important industries today are the metallurgical and engineering industries, and shipbuilding and ship repairing.

*Rotterdam* (1,031,000) developed on a small tributary of the Lek (the main distributary of the Rhine), and has a large and productive hinterland covering the whole Rhine basin, including the Ruhr industrial region. Today its link with the sea is by the 29 kilometre-long New Waterway. This roughly follows an old distributary, but is really a new canal, finished in 1872 and subsequently enlarged several times to keep pace with the growing volume and size of shipping.

Rotterdam is the world's greatest port. Two-thirds of its trade is transit trade, which includes bulk imports of petroleum, iron ore, raw materials and foodstuffs much of which are sent up the Rhine by barge or small steamer to Germany, France and Switzerland, and exports of coal, potash and manufactured goods. In recent years the

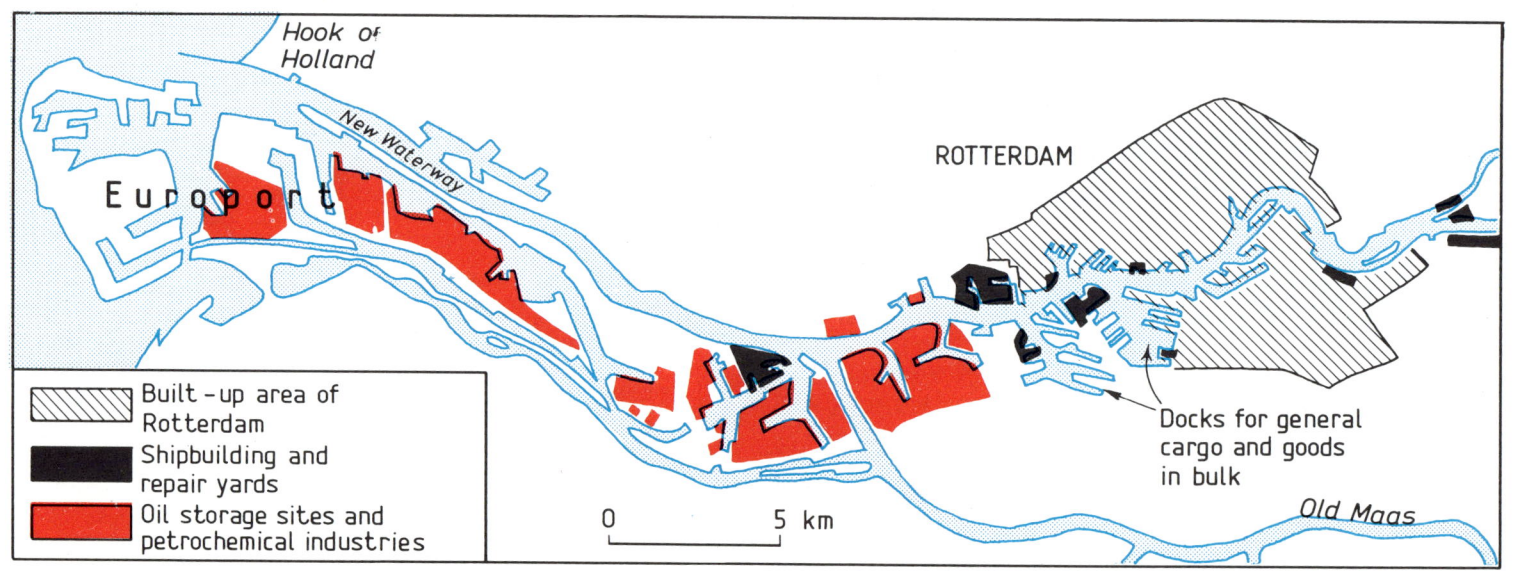

FIG. 6.8 **The Port of Rotterdam**

**Rotterdam**

*J. Allan Cash*

Note the evidence of transit trade by road, river and railway.

**Europort**     *Dutch Embassy*
Oil tankers at Verolme shipyards.

dock facilities have been reorganised, and since 1958 vast new dock systems have been built at Europort at the end of the New Waterway opposite the Hook of Holland, to serve as an international port for the Common Market countries. There has been great industrial expansion especially along the New Waterway where there are shipyards, engineering works, soap and margarine works, and oil refineries. A ferry service operates from the Hook of Holland to Harwich.

*The Hague* (680,000) is the seat of government and the royal residence. It is mainly a residential city, with attractive avenues and squares, but there are also many light industries including engineering, clothing, food processing and printing.

The other large towns are primarily engaged in manufacturing, but usually combine this with their function as regional administrative centres and market towns. Many of these centres are concerned with general manufacturing, including various branches of the engineering, textile and chemical industries. *Utrecht, Arnhem, Haarlem* and *Nijmegen* all belong to this group. Other towns have more specialised industries, for example *Tilburg* and *Enschede* are centres for textile manufacturing, and *Eindhoven* is the home of the giant Philips electrical works.

## Exercises

*Answer in note form:*

1. Describe the methods of reclamation of (a) the polders and (b) the eastern heathlands.
2. What is the purpose of the 'Delta Plan'?
3. Explain why Rotterdam has developed into the world's largest port.

*Essay Question*

What geographical factors have encouraged the growth of manufacturing industries in the Netherlands in the post-war years? Describe briefly the character of industry today, mentioning the main centres of production.

# 7: BELGIUM AND LUXEMBOURG

Belgium has an area of 30,600 square kilometres and a population of nearly 10 million, giving a density of nearly 330 to the square kilometre. This makes it, after Holland, the second most densely populated country in the world.

National unity has been achieved despite the fact that there are two distinct communities, the mainly Protestant Flemings in the west and north, who speak a language which is akin to Dutch, and the Roman Catholic Walloons in the south, who speak a French dialect. The country lacks natural frontiers and is a 'buffer state' between France and Germany. As a result, it suffered severely during the two world wars, but has recovered quickly. Since the last war there has been a rapid growth of industry and trade, aided by membership of the Benelux union and the European Economic Community.

Belgium has well-defined physical regions, the main contrast being between the dissected Hercynian upland of the Ardennes in the south-east, and a broad lowland in the north-west (Figure 7.1). The lowland may be sub-divided into the Sambre-Meuse trough, the rolling Mid-Belgian Plain, and the lower Flanders Plain. The latter includes the flood plains of the Scheldt and Lys, and a coastal zone with a line of sand dunes on the seaward side, backed by a narrow belt of polder country. In the eastern part of the lowland is the sandy Campine which has, until the present century, been left as poor heath and woodland.

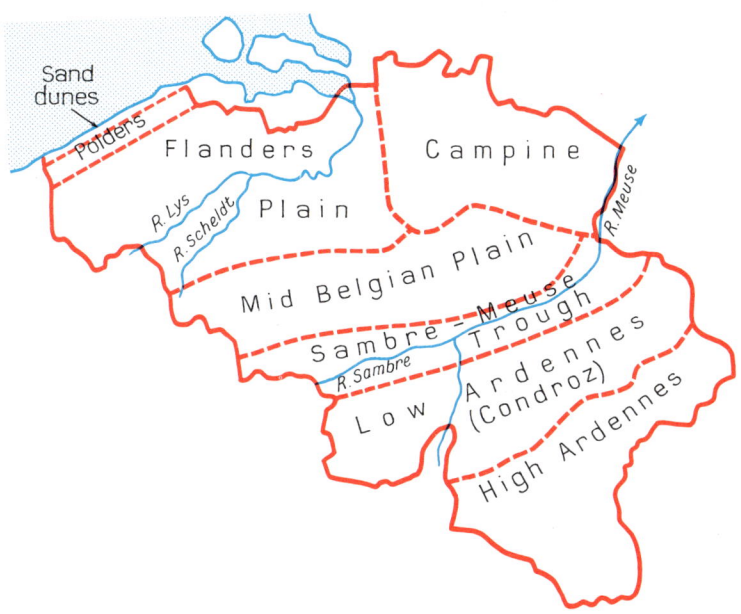

FIG. 7.1 **The regions of Belgium**

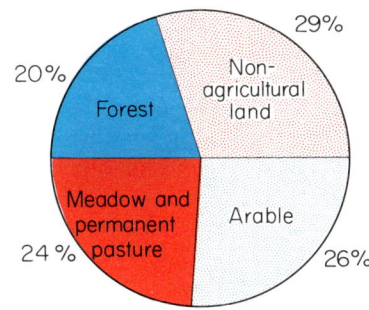

FIG. 7.2 **Belgium—land use**

Climatic conditions are similar to those of the Netherlands. Maritime influences predominate, producing mild winters and cool summers, with a moderate, well-distributed rainfall. Brussels, whose climate is typical of the northern lowland, has an average annual rainfall of 760 mm, and mean monthly temperatures ranging from 2°C in January to 18°C in July. The rainfall is heavier and temperatures considerably lower in the Ardennes.

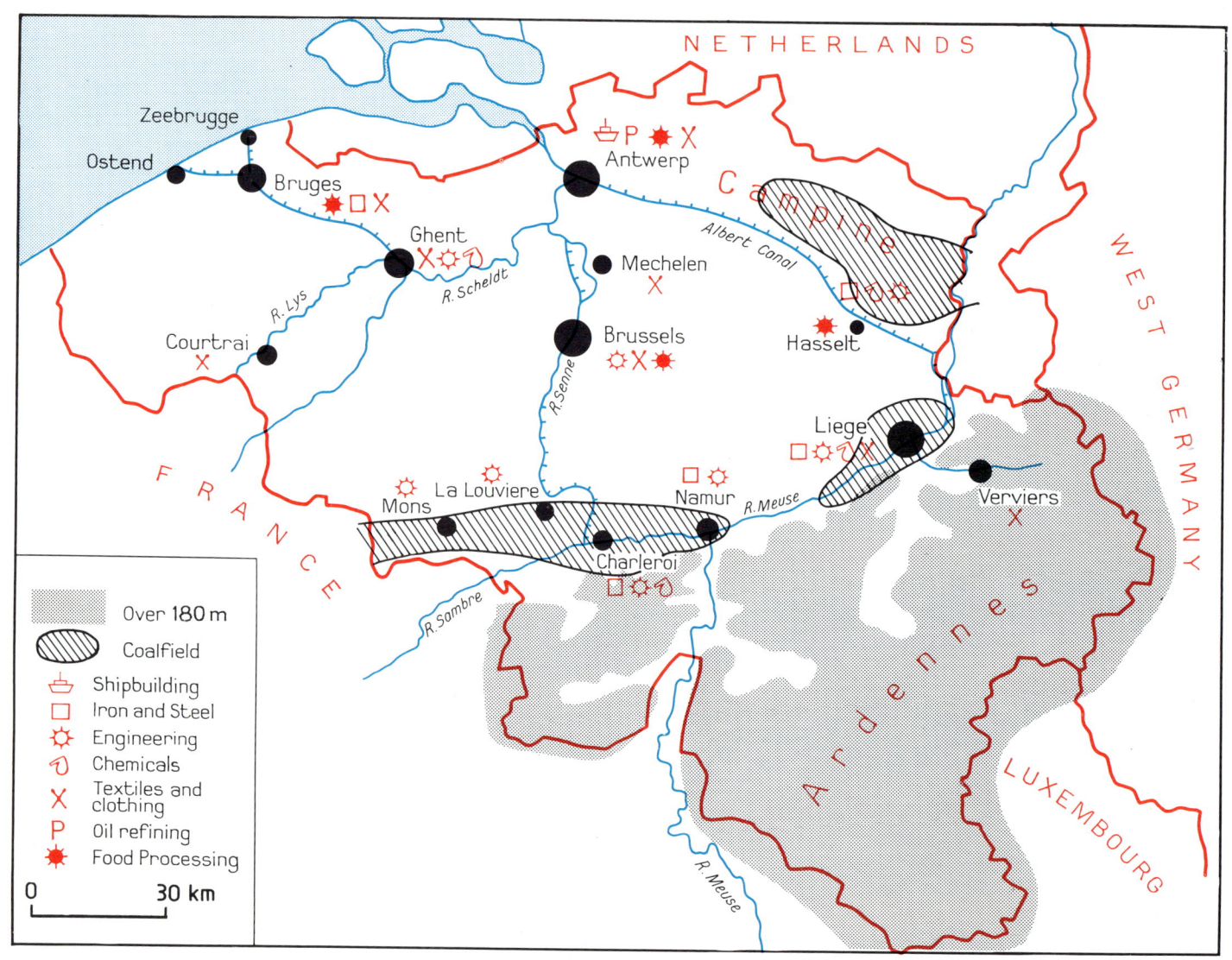

FIG. 7.3 **Belgium**

## Agriculture

Agriculture employs only 3·5 per cent of the working population, but supplies 90 per cent of the temperate foodstuffs needed by the country. The average farm covers only about 13 hectares; almost all the labour is supplied by the farmer and his family. Most farmers grow a variety of crops and keep cattle, pigs and poultry. Methods are among the most intensive in the world, and large quantities of fertiliser are used, but much of the work is still done by hand. Efforts are being made to persuade the older farmers who have very small holdings, and the factory workers who do some farming in their spare time, to sell their land so that larger, more efficient farms can be created which would give more scope for the use of modern machinery.

## Industry

Belgium is mainly a manufacturing country. Industry has been handicapped by the relatively poor natural resources and the need to import most of the necessary raw materials. On the other hand it has been helped by its location in the centre of one of the most concentrated industrial areas in the world, stretching from the north French coalfield through Belgium to the Ruhr and from Lorraine to the Rhine estuary, and also by an excellent system of internal communications by rail, road and waterway.

Coal is the only significant raw material although production from the Sambre-Meuse coalfield, which is where most of the heavy industry is located, has greatly declined in recent years. At one time, at the end of the nineteenth century, there were over 250 collieries; now there are only 10 but these are large, highly organised pits which together produce some 7 million tonnes of coal per annum. Of this, 85 per cent comes from the Campine field. In the past iron ore as well as lead and zinc have been mined, and while these resources played a part in establishing early industries, production today is negligible.

The histories of both textile production and metal-working go back to the Middle Ages, and the acquired skills have been passed on from generation to generation. The traditional textile industry of the Flanders Plain has continued although it is much smaller. Today much of the production is concerned with synthetic rather than natural fibres. The heavy iron and steel and chemical industries are located in the Sambre-Meuse area, but engineering and chemical industries have recently been established on the Campine coalfield.

The growth of industry during the period 1957–1967 has been remarkable; production in the petro-chemical industries has doubled,

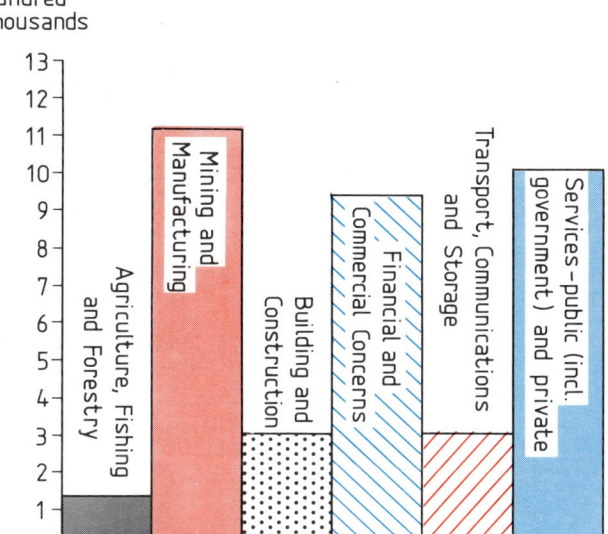

FIG. 7.4 **Main sectors of employment in Belgium**
Comment on the differences between this figure and Fig. 6.4.

and there have been great increases in the output of machinery and non-ferrous metals. Forty per cent of the industrial production, including metal and engineering products, textiles, chemicals and glassware, is exported, chiefly to other West European countries.

# THE REGIONS OF BELGIUM

## The Ardennes

The massif of the Ardennes, which consists of old, hard rocks, is structurally a horst, i.e. a block of land which has been faulted round the edges and then raised to form a plateau. It reaches 600 metres towards the south, and has been deeply dissected by rivers. The heavy rainfall, impervious rocks and thin, acid soils make it an infertile area. The rounded summits are mostly moorland and the slopes are forested with pine and spruce, but also contain clearings for cattle pasture. There is some employment in forestry, the timber being used for pit props and constructional work, and the region is visited

**The flanks of the High Ardennes** *Belgian National Tourist Office*

**A village on the edge of the High Ardennes** *J. Allan Cash*

by large numbers of tourists. In the extreme south-east younger rocks extend into Belgium from French Lorraine. The working of iron ore in these rocks led to the development of an iron and steel industry, and this is still carried on in a number of small towns, although only one iron mine is now active.

To the north-west lies the Low Ardennes or Condroz. Here the land is below 300 metres and has a less severe climate, with more fertile, loamy soils which have developed on limestone rocks. Whilst con-siderable areas are still forested, this is a significant dairying region supplying milk to the large industrial population immediately to the north.

## The Sambre-Meuse Depression

The valleys of the Sambre and Meuse extend along the northern edge of the Ardennes. There were scattered industries here before the

**A farming scene in the Condroz** *Institut Belge d'Information et de Documentation*

**The Meuse Valley near Liege**

*Institut Belge d'Information et de Documentation*

Industrial Revolution including metal-working, using local supplies of iron and zinc, and charcoal from the Ardennes. Large-scale expansion began in the early nineteenth century, and today the main industrial areas coincide approximately with the workable deposits of coal. Although the coalfield has numerous seams so that it is possible to mine many from one shaft, the seams are difficult to work since they are thin and contorted by folding and faulting. In recent years, since the coal mining industry has declined to such an extent, there has been a need to attract new industry to the area. There are new industrial estates with light industry around Mons and Charleroi.

In spite of the fact that local iron ore deposits are exhausted and there is insufficient coking coal, the Sambre-Meuse corridor remains a major steel producing area and there are new steel works at Liège. Iron ore is imported mainly from Algeria, Sweden and Lorraine, and coke is brought in from the Ruhr. This tendency of an industry to stay where it grew up, even when the original factors which led to its growth no longer apply, is known as industrial inertia. There are many other examples in North-West Europe.

Apart from coal mining and steel, there is heavy engineering, zinc and copper-working, and the manufacture of chemicals and glass. The main industrial towns are *Mons*, *Charleroi*, *Namur* and the largest, *Liège*, which with its industrial suburbs has a population of half a million. The growth of Liège has been helped by its position at the junction of major routeways, not least the motorways which radiate from it. It has a number of recently developed industrial suburbs especially to the north of the city. The Sambre and Meuse form an important line of communication; the rivers are canalised and used for transporting coal, iron ore and other bulky goods, and are followed by a main road and railway. Liège can be reached by sea-going ships from Antwerp via the Albert Canal.

In the east, on a small tributary of the Meuse, is *Verviers*, which has a long-established woollen industry originally based on water power and local supplies of wool.

## The Mid-Belgian Plain

This is a region of rolling, low plateaus and broad valleys. It consists mainly of sands, chalk and clays, but nearly everywhere these are masked by a thick covering of loess. This produces rather bare country under intensive agriculture, particularly the growing of wheat and sugar beet, with orchards and market gardens giving variety to the scene. There are large areas of glasshouses near Brussels, well known for the production of table grapes.

*Brussels* (1,075,000 including suburbs) occupies a central position in the country and is a major focus of road and rail routes. There is a canal linking it with the Scheldt, which can be used by small sea-going vessels, and another to Charleroi. Brussel's new road systems, its modern shops and offices, and many beautiful old buildings, make it an impressive capital. It is the country's chief commercial and industrial centre, its activities including printing, the manufacture of high quality clothing and many other types of consumer goods, and heavy engineering. It is the headquarters of the European Economic Community.

**Brussels, the Grand Place and Town Hall**    *J. Allan Cash*

**The Campine**    *Belgian National Tourist Office*

## The Campine

The Campine is a low plateau of infertile glacial outwash material. It has been, like much of the sandy plateau of the Netherlands, a zone of heath and marshland, but in recent years there has been extensive reclamation, involving the drainage of the marsh and heavy manuring of the sand. Dairying is the most important aspect of farming and near Antwerp market gardening has been encouraged by the nearness of a large industrial population. On some of the poorer land pine plantations have been developed.

It is in the Campine that the country's most important coalfield is located. This is a continuation of the Limburg field in the Netherlands, and its development dates only from 1917. Mining is carried on at a number of large collieries and although the coal is at great depth, the seams are easier to work than in the Sambre-Meuse coalfield. The coal has encouraged the establishment of new industries, including metal-working, engineering and chemical manufacturing, particularly along the Albert Canal. The availability of cheap land and easy access to the port of Antwerp are other factors favouring industrial development.

54

# The Plain of Flanders

Flanders is a region of varied soils which have developed mainly on clays and sands, and includes the alluvial plains of the Lys and Scheldt. Heavy manuring and intensive farming have made it a productive agricultural area. The principal crops are wheat, sugar beet, hemp, flax and potatoes, and large numbers of dairy cattle and pigs are kept. The narrow coastal plain consists of drained polders with heavy clay soils mainly under grass for dairy cattle, and pockets of sand and peat where arable farming has developed. The coast itself is an almost continuous line of sand dunes, with numerous holiday resorts. Marram grass and belts of pines have been planted to prevent the sand from drifting inland.

The Flanders Plain has been a textile manufacturing region since the Middle Ages. The woollen and linen industries developed first, making use of local wool and flax, and soft river water for the washing and retting processes (retting is the process whereby the flax is steeped in water so that the fibres can be more easily separated). During the nineteenth century cotton replaced woollen manufacturing, and in recent years increasing use has been made of synthetic fibres, including rayon and nylon. Many towns have been concerned with textile manufacturing but the main centre is *Ghent*. This fortress town grew up on an island at the confluence of the Lys and Scheldt, and has clothing, leather and furniture industries as well as textiles. *Bruges* is a historic town whose ancient buildings attract many tourists. Lace-making and other hand industries are still carried on, but it also has iron and steel and engineering industries.

### The Polder Zone near Zeebrugge

*Belgian National Tourist Office*

Study the photographs in this chapter and then describe each (except for the Brussels view) pointing out the main geographical distinctions of the regions they illustrate.

**The Dune Coast**  *Institut Belge d'Information et de Documentation*

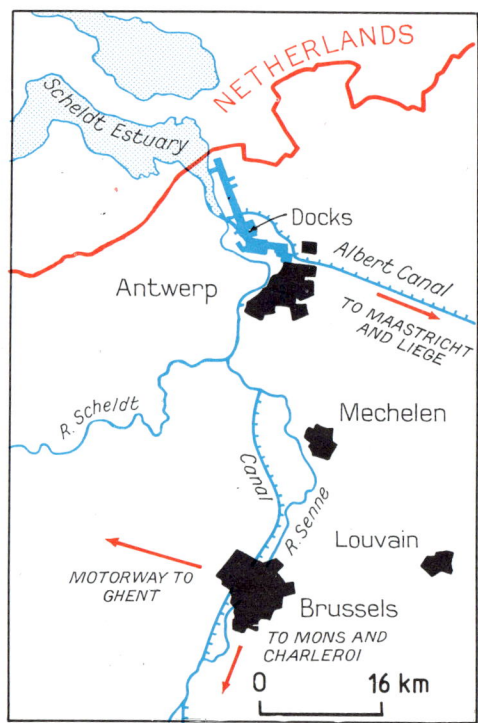

FIG. 7.5 **The position of Antwerp and Brussels**

A ship canal connects Ghent with Bruges and the commercial port of *Zeebrugge* on the coast. *Ostend* is a holiday resort and fishing port, and has a ferry service to Dover.

*Antwerp* (700,000) lies on the right bank of the Scheldt estuary, about 80 km from the sea. It is a natural focus of routes not only from much of Belgium, but also from parts of north-east France and West Germany. Thus it is a major European port, and handles most of the trade of Belgium and Luxembourg. One feature is the large amount of barge traffic, especially to and from Liège and Brussels.

Many industries have developed in the Antwerp area. Along the estuary are shipyards, oil refineries, cement and brick works, and factories processing imported foodstuffs. In the city itself and in the surrounding industrial suburbs are heavy engineering works, tobacco and clothing factories, and numerous other light industries.

From Antwerp's southern industrial suburbs and extending towards both Ghent and Brussels a major industrial triangle has developed in which engineering, car assembly, synthetic textile and plastics plants are all important.

## THE GRAND DUCHY OF LUXEMBOURG

This small principality lies between Belgium, France and Germany and has a population of some 360,000. It shows many French and German influences, both languages being widely used, but its economic associations have recently been more closely linked with the Benelux union alongside Belgium and the Netherlands.

The country can be divided into two distinct parts, the Ardennes upland and the southern scarplands. The Ardennes occupy the northern part of Luxembourg and are a moorland plateau with thickly wooded slopes. The lower slopes and interior valleys have long been concerned with cattle rearing together with the cultivation of oats and potatoes. Farming methods have improved in recent years and dairying has extended here.

The southern scarplands, which are often called the 'Bon Pays', consist of younger sedimentary rocks and have better soils than the Ardennes. Much of the land is under mixed farming, most farms having a small herd of dairy or beef cattle, whilst the arable land is mainly used to produce fodder crops such as oats and potatoes for the livestock. There are orchards and vineyards on hill slopes, especially on the south-facing slopes overlooking the Moselle valley which have a particularly mild, sunny climate.

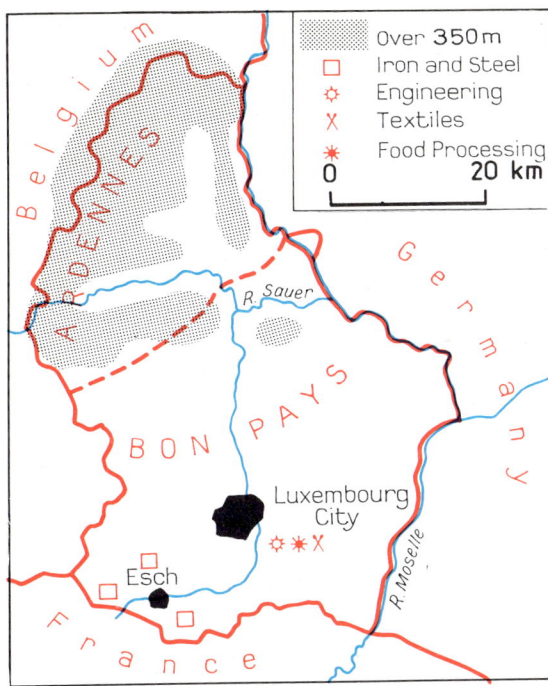

FIG. 7.6 **Luxembourg**

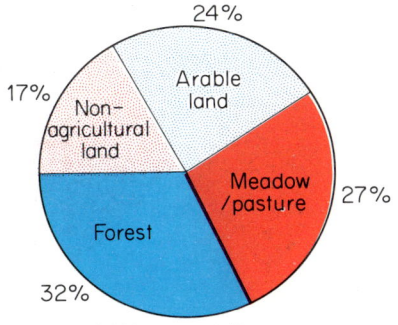

FIG. 7.7 **Luxembourg—land use**

The basis of Luxembourg's economy is its flourishing iron and steel industry which developed because of iron ore deposits that form an extension of the Lorraine ore field into the country. Coke is imported from West Germany, and iron ore is brought in from Lorraine and even Sweden, as the domestic ores are becoming worked out. Steel production exceeds 4 million tonnes, and most of the steelworks lie south of Luxembourg city, close to the border with Lorraine. Most of the steel is exported, but some is used in local engineering industries. These and other industries including textiles, hosiery and tanning, have developed in the city of *Luxembourg*, which grew up on an easily-defended site at the meeting place of routes from France, Germany (via the Moselle) and Liège. Luxembourg City is also a significant commercial and administrative centre of Western Europe, with banking concerns and some of the offices of the European Economic Commission.

## Exercises

*Answer in note form:*

1. What are the reasons for the lack of population in the Ardennes and the dense population along the Sambre-Meuse valley?
2. Why is the textile industry located mainly in the Ghent area?
3. In what ways do the functions of Antwerp and Brussels differ, and in what ways are they similar?

*Essay Question*

Make a careful copy of this section, printing in the names of the regions represented by the letters A, B, C, D, E, F and G.

Explain how the soils and underlying rocks have influenced the character of the farming along the line of the section.

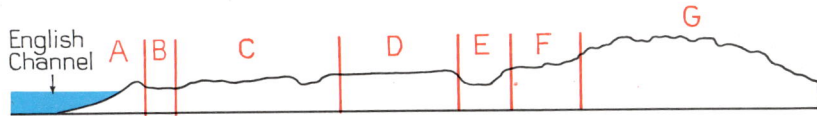

FIG. 7.8 **Section across Belgium from the north-west to the south-east**

# 8: FRANCE

France has a population of 53 million and an area of 553,000 square km, making it the largest of the west European nations. It has a moderate density of population, 94 per square kilometre, and a well-balanced economy. Agriculturally the country is self-supporting in basic foodstuffs, and industry has developed greatly during the last twenty years. The strength of the economy lies in the variety of physical regions and climates, and the abundant natural resources. Membership of the European Economic Community has further strengthened France's position as a trading nation.

## Structure and Relief

France is a land of great physical variety and there are three main elements in its structure. Firstly there are rugged fold mountains formed during the Alpine mountain-building period, including the Alps themselves, as well as the Jura and Pyrenees. Secondly there are plateaus of old, hard rocks, worn-down remnants of the former Hercynian fold mountains, of which the Massif Central, Armorican Peninsulas and Vosges are the main examples. Thirdly there are lowland basins floored by young sedimentary rocks, the largest being the Paris Basin, which is drained by the Seine and middle course of the Loire, and others being the Aquitaine Basin, drained by the Garonne, and the much narrower trough formed by the valleys of the Rhône and Saône.

## Climate

The climate varies considerably in different parts of the country. The north and west are under mainly maritime influences with moist, equable air streams from the Atlantic, giving cool summers, mild winters and moderate precipitation falling all the year round. Eastern France and the Massif Central have a more continental type of climate with colder winters, warmer summers (on the lower land), and more rain in summer than in winter. Along the south coast the climate is of Mediterranean type with warm, dry, sunny summers and mild, rainy winters. In mountainous regions, particularly the Alps, Pyrenees and the upper parts of the Massif Central, conditions are greatly modified by the altitude, temperatures being markedly lower and precipitation greater, with heavy winter snowfall.

## Agriculture and Fishing

Because of the large areas of fertile soils in its great river basins and the moderate population density, France is able to produce almost all the food it needs. At present about 11 per cent of the working population are employed in agriculture, which is a large proportion for a west European country, but the number of farm workers is steadily decreasing. Forty per cent of the holdings are of less than 10 hectares; many are too small to be efficiently worked. A considerable number are fragmented, which means that one farmer can have many small and entirely separate pieces of land to work. The government policy of remembrement, particularly in the west of France, is the amalgamation and regrouping of farms to make them more viable.

Much of the arable land is concentrated in the lowlands or on the limestone and chalk plateaus. Cereals are the most important crops, with wheat outstanding especially in the Paris Basin and north-east, followed by oats and barley. Maize is largely confined to the Aquitaine Basin, whilst Brittany and the Massif Central have the greatest acreages of potatoes. Fruit production is remarkable for its variety, ranging from apples and pears in the north-west to peaches in the Rhone valley, whilst vineyards are found in many parts of the country.

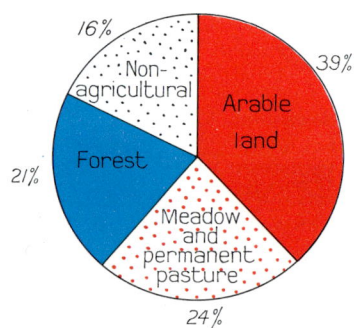

FIG. 8.1 **France—land use**

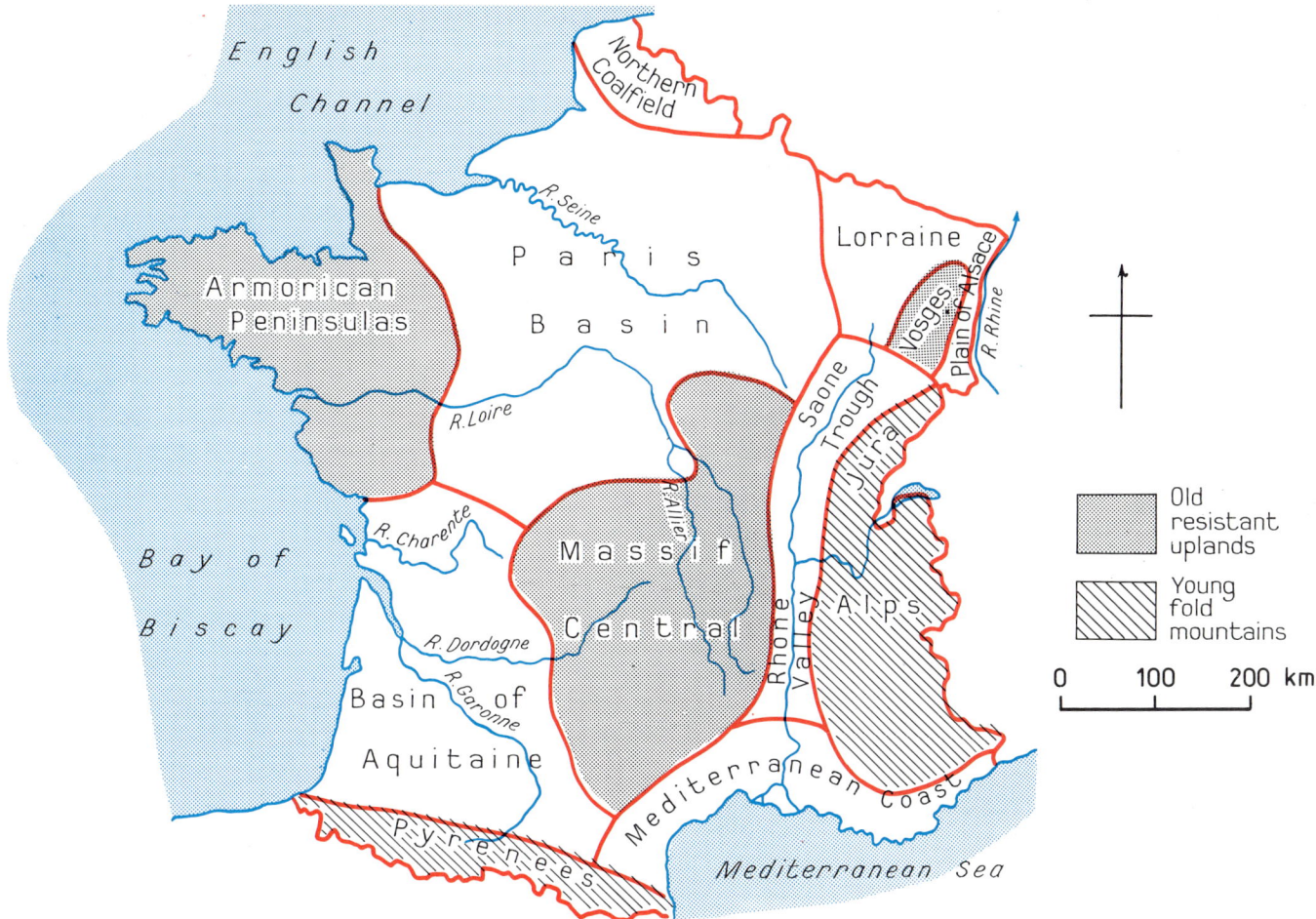

FIG. 8.2 **The regions of France**

The uplands have the most extensive pastoral farming areas with large numbers of cattle in Armorica, the Vosges and Massif Central. France is a major producer of beef, exporting considerable quantities to West Germany. Sheep are declining in number.

France has an old-established fishing industry, with a number of trawler fleets operating from the Channel and Breton coasts. Many Breton fishermen are part-timers and use small-sized sailing boats.

On the whole the industry is less well organised than in other North-West European countries.

### Industry

In industrial production, France has been overtaken by West Germany. One reason for this is the poor quality of French coal, and the difficulties of mining it. Coal production has declined markedly in recent years, particularly in the six years between 1967 and 1973 during which time annual production was halved. Half of the present total production of some 21 million tonnes comes from the Moselle coalfield in Lorraine. The Nord coalfield and the pockets of coal around the Massif Central are now of minor importance.

France is relying increasingly on other sources of energy. At present half of its electricity is derived from hydro-electric stations in the Alps, Massif Central and Pyrenees. Large quantities of natural gas are obtained from the foothills of the Pyrenees, but domestic oil production, again from the south-west, provides only a small proportion of the country's needs. There are plans for a rapid expansion in the production of nuclear energy, and even the tides are used to operate a power station in the Rance estuary in Brittany.

Among the other natural resources are vast deposits of iron ore in Lorraine, and smaller ones in Normandy, the Pyrenees and Massif Central. One third of the total production is exported, mainly to

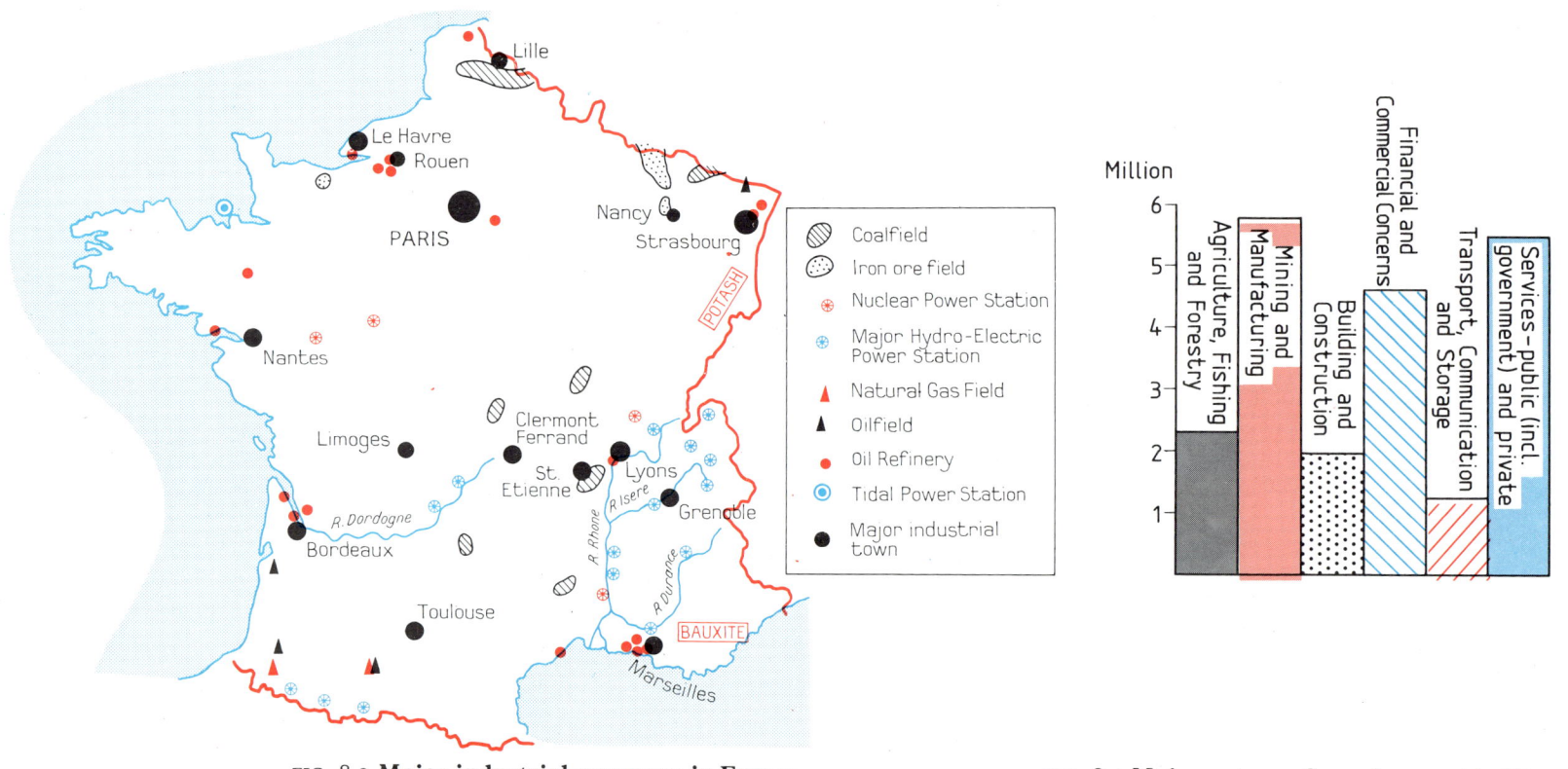

FIG. 8.3 **Major industrial resources in France**

FIG. 8.4 **Main sectors of employment in France**

Belgium, Luxembourg, and the Saar in West Germany, but being low-grade ore, it is not in such great demand as formerly, and production is declining. Apart from iron ore there are important bauxite deposits in the south Alpine region, and potash in Alsace.

The main branches of manufacturing are all well-represented in France. The iron and steel industry is concentrated on the Lorraine ore field and the Nord including the coalfield area and the coast, with new coastal works at Fos near Marseilles, and at Dunkerque. Engineering is very widely dispersed. The main textile area is on or near the Nord coalfield, with smaller developments in eastern France and along the Rhône valley. As with so many industrial countries, the most spectacular growth has been in the chemical industry whose output has increased fourfold in the last ten years, plastics and fertilisers being the two leading contributors to this growth.

The dominant feature of French industry is the great diversity of manufacturing with new industrial estates on the outskirts of all major towns, and the tendency now is for expansion in the predominantly rural western half of the country helped by government policy.

## THE REGIONS OF FRANCE

### The Armorican Peninsulas

Armorica consists of Brittany, the Cotentin peninsula of Normandy, and the lower Loire valley. Much of the land is a low plateau rising to between 200 and 400 metres above sea level, with the high ground composed of granite and metamorphic rocks, and the lowlands of shale. In Brittany the land drops steeply to a rugged, indented coastline containing inlets (rias) caused by the drowning of the lower courses of the rivers.

The climate is remarkably equable, with very mild winters and cool summers. The rainfall is evenly distributed and totals about 840 mm on the lowest land and considerably more on the hills. The almost frostless winters and early arrival of spring along the coast are great assets to agriculture.

The coastal areas are the most densely populated. Market gardening is important especially on the north Brittany coast where for centuries the land has been fertilised by seaweed and crushed shells. Here there is a concentration on the production of early vegetables, flowers and small fruits such as strawberries. There are also apple orchards, much of the crop being used for the production of cider.

Dairying is expanding and large quantities of milk are made into butter and cheese.

The highlands of the interior remain infertile moorland, but the interior lowlands which extend well into Normandy and Maine have a characteristic scenery which the French call 'bocage' (meaning wooded). Tall hedgerows grow from earth banks between small fields, and these together with the numerous small copses give a well-wooded appearance to the land. Farming here is mainly concerned with livestock, including the rearing and fattening of cattle and dairying,

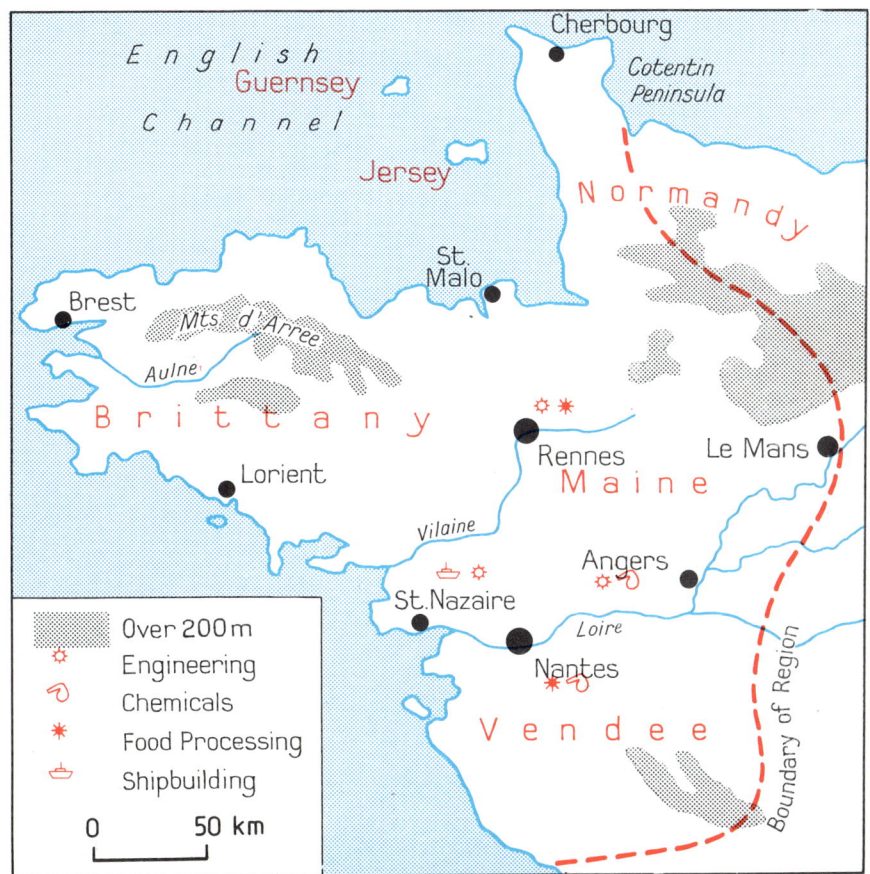

FIG. 8.5 **Armorica**

whilst some areas have a long tradition for the rearing of horses. Most of the arable land is used for growing fodder crops.

This is the most important fishing area in France. There is inshore fishing for sardines, lobsters and tunny in south Brittany, whilst in the north the catch is more varied and includes turbot and mullet. The north coast also is associated with deep sea fishing and fleets from here fish for cod in Icelandic, Arctic and Newfoundland waters.

There are many ports and holiday resorts round the coast. *Brest* lies on a deep, sheltered inlet in the west of Brittany, but its remoteness from large industrial areas has restricted its growth and it is mainly a naval base. *Lorient* is a leading fishing base, *Cherbourg*, in the north of the Cotentin peninsula, is a port of call for Atlantic liners, and *St Malo* is a commercial port which exports farm produce to Britain. There are ferry services from both Cherbourg and St Malo to Southampton. *Nantes* is an important port at the head of the Loire estuary, but because of silting it is linked to the sea by a canal. Its main industries are the processing of bulk imports including flour milling, sugar refining and soap production, whilst its outport, *St Nazaire*, has shipbuilding and marine engineering industries. The largest inland town is *Rennes*, the capital of Brittany and a market town serving a rich agricultural area, with food processing and new engineering industries.

**Citroen Works near Rennes**  *Photo Citroen*

**Pointe du Raz, Brittany**  *Photo Viguier*
The granite ridges end in rugged headlands, which are under constant attack by the waves.

## The Aquitaine Basin

The Aquitaine Basin is a triangular-shaped lowland bordered by the Massif Central in the east and the Pyrenees in the south, and linked to the Paris Basin by the Col de Poitou and to the Mediterranean by the Col de Carcassonne. It is primarily an agricultural region with only two large industrial towns (Toulouse and Bordeaux), although in recent years industry has developed in the south, influenced to some extent by hydro-electric developments in the Pyrenees and local natural gas and oil.

The basin is made up of Tertiary sands and clays, with limestones along the eastern fringes. The smooth coastline is the result of deposition by the longshore drift, and in a number of places lagoons have been cut off from the sea by sand bars.

The climate is essentially of maritime type with mild winters, as in Brittany, but warmer summers because of the more southerly latitude. The rainfall, of about 750 mm, is well-distributed throughout the year, although it is slightly drier in summer.

The Garonne valley is the heart of Aquitaine and is noted for its rich and varied pattern of agriculture. More than half the land is under cereals, mainly wheat and maize, and there are market gardens, orchards and vineyards especially on the river terraces, and cattle pastures on the valley floor. The estuary area and the Medoc peninsula are particularly noted for their vineyards, which produce the high quality 'vins de Bordeaux'. The smaller Charente valley is another rich farming area with much land under wheat, orchards, vineyards and dairying, and is noted for the production of the world-famous Cognac brandy, named after the small town of Cognac. In the east of the region are low, bare chalk and limestone plateaus, mainly used for sheep rearing.

There are two infertile areas in the Aquitaine Basin. The first is the *Landes* in the west, which has been invaded by sand blown inland from the coastal dunes. These sands are underlain by a hard pan cemented by iron oxide, so that the land is poorly drained. Reclamation has been taking place for more than a hundred years. The coastal dunes have been fixed by the planting of deep-rooted grasses, and in the interior much of the land has been drained and planted with pine forests which produce valuable supplies of timber and turpentine. The second infertile area is in the south where immense fan-shaped deposits of gravel have been brought down by rivers from the Pyrenees. These form bare plateaus which provide poor pasture for sheep, but there is better land along the rivers which have cut deeply into the plateaus.

The Aquitaine Basin is a popular holiday area and the sandy beaches have led to the growth of many coast resorts, the largest of which is *Biarritz*. The main outlet for the region is *Bordeaux* (555,000), which grew up at the lowest bridging point of the Garonne, some 100 km from the sea, and is the highest place that can be reached by sea-going ships. It is an important commercial centre dealing particularly with the wine and timber trades, and many of its industries are concerned with the processing of imports, including sugar refining, the pressing of oil seeds, and oil refining. In recent years there has been a considerable expansion of other industries, notably engineering and chemicals.

*Toulouse* (450,000) is the chief industrial town, and receives hydro-electric power and natural gas from the Pyrenees. Its manufactures are of great variety, ranging from food processing to the aircraft and chemical industries. Its growth owes much to its position at a focus of routes and its command of the important Col de Carcassonne.

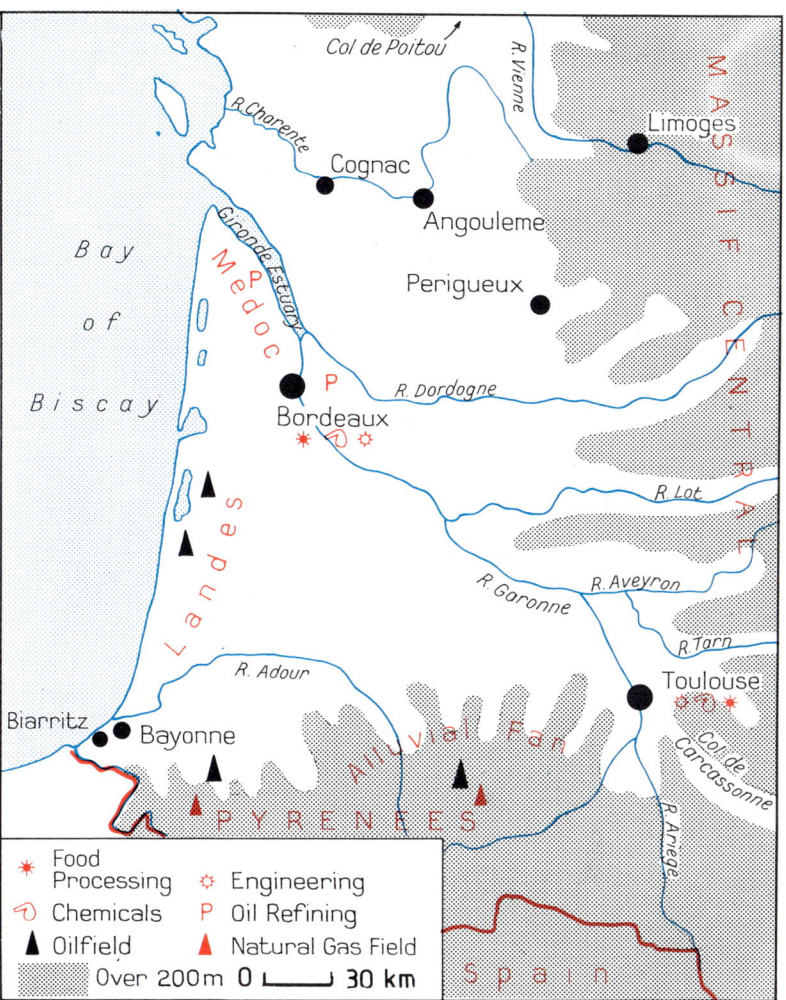

FIG. 8.6 **Aquitaine**

## The Pyrenees

The Pyrenees are a range of rugged fold mountains with a very steep northern face, and form a formidable frontier between France and Spain. The easiest routes pass round the western and eastern ends, but there are a number of high passes, two of which are used by railways.

In the west, where the mountains reach only about 900 m, there is quite a heavy rainfall of between 1,000 and 1,500 mm per annum, and the slopes are well wooded with deciduous trees. Soil erosion is a serious problem where woodland has been cleared, especially from the steeper slopes. There is some good farmland on the lower slopes and in the valleys, where orchards, dairying and cereals are all important.

In the central Pyrenees, which reach 3,404 m in the Pic d'Aneto (in Spain), there are still some small glaciers and many features of former glaciation including vast cirques. The deep, narrow valleys have in many places been dammed for the production of hydro-electric power. The schemes are small but numerous and have led to some local industrial development especially aluminium refining, the making of special steels in electrical furnaces, and a chemical industry. In the foothills is the famous pilgrimage town of *Lourdes*, and also a number of important oil and natural gas fields.

The lower eastern Pyrenees have less rain as a result of the influence of the Mediterranean climate. Evergreen oaks and maquis-type shrubs are typical here, and on the lower slopes where the land is terraced and irrigated, fruit, vines and vegetables are grown, but elsewhere sheep and goats are kept on the poor pasture. *Perpignan*, in the foothills, has varied industries including the manufacture of paper, leather and textiles.

**Natural gas field at Lacq**  *French Government Tourist Office*

Natural gas was discovered at Lacq, in the foothills of the Pyrenees, in 1949 and present annual production is equivalent in heating power to 6 million tonnes of coal. The gas is carried by pipeline to Paris and Bordeaux.

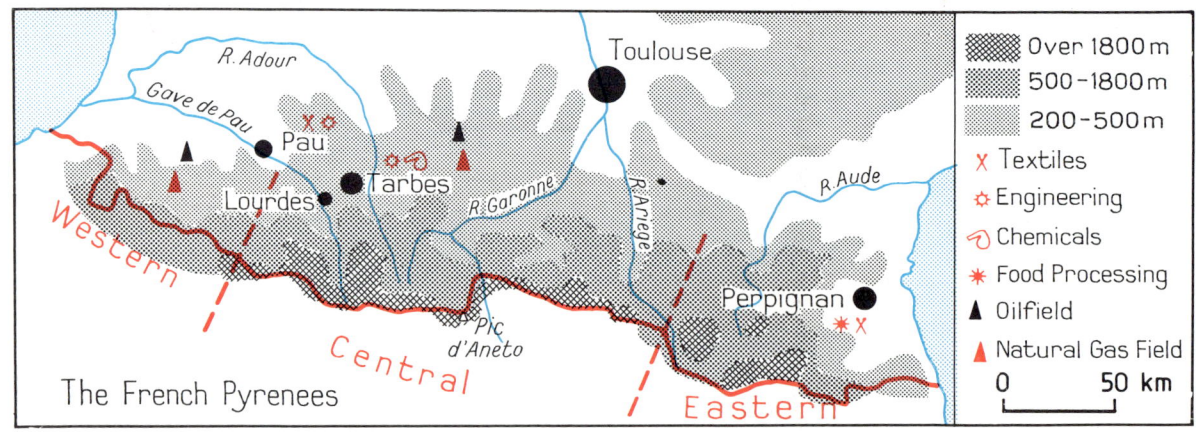

FIG. 8.7 **The French Pyrenees**

## The Massif Central

The Massif Central is a plateau composed of resistant rocks, with an average height of 1,000 metres and reaching 1,800 metres in places. The rocks were folded in the Hercynian mountain-building period, then worn down to a peneplain, and later uplifted to form a tilted block which slopes down towards the west and north. There have been widespread igneous intrusions and volcanic activity, the latter mainly in the Auvergne where the plugs of extinct volcanoes as well as lava plateaus can still be seen.

Altitude gives the region a cool, damp climate, with an annual precipitation in excess of 1,140 mm, including considerable winter snowfall on the higher parts. The scenery is closely related to the rock type and structure, and five distinct sub-regions can be recognised.

In the north-west is the almost featureless granite plateau of *Limousin*. The thin, infertile soils of the higher land are covered with heath and forest, but on the lower western edge the soils are better, and there is cattle rearing and the growing of hardy crops including rye, oats, buckwheat and potatoes. *Limoges* is the principal town of the region; it is a big market centre and has a variety of specialised industries of which the manufacture of porcelain and china, based on local kaolin, is a long-established one. In recent years, after a long period of stagnation, new industries including the car industry have revitalised the town.

In the south and south-west the plateau is formed partly of crystalline rocks and partly of limestone. The limestone areas are known as *Causses* and have a barren karstic landscape, the surface being cut into by deep gorges and pitted with depressions. Many streams flow underground but there are a few surface ones including the Tarn which flows in a spectacular gorge whose sides rise almost vertically for 600 metres. The Causses provide poor grazing for sheep, whose milk is used to make the famous Roquefort cheese, which is matured in caves in the limestone. It is a region in which depopulation has been very marked. The crystalline areas are more productive and their soils have been improved by the application of lime and fertilisers, so that large numbers of cattle are kept, and wheat, potatoes and fodder crops are grown.

The most impressive scenery is in the *Auvergne*, where a line of extinct volcanic cones known as puys can be traced, the highest being the Puy de Dome, 1,465 m. Farther south lie the great volcanic masses of Mont Dore (which includes the peak of the Puy de Sancy, 1,886 m) and the dome of Cantal. There are basalt lava flows which

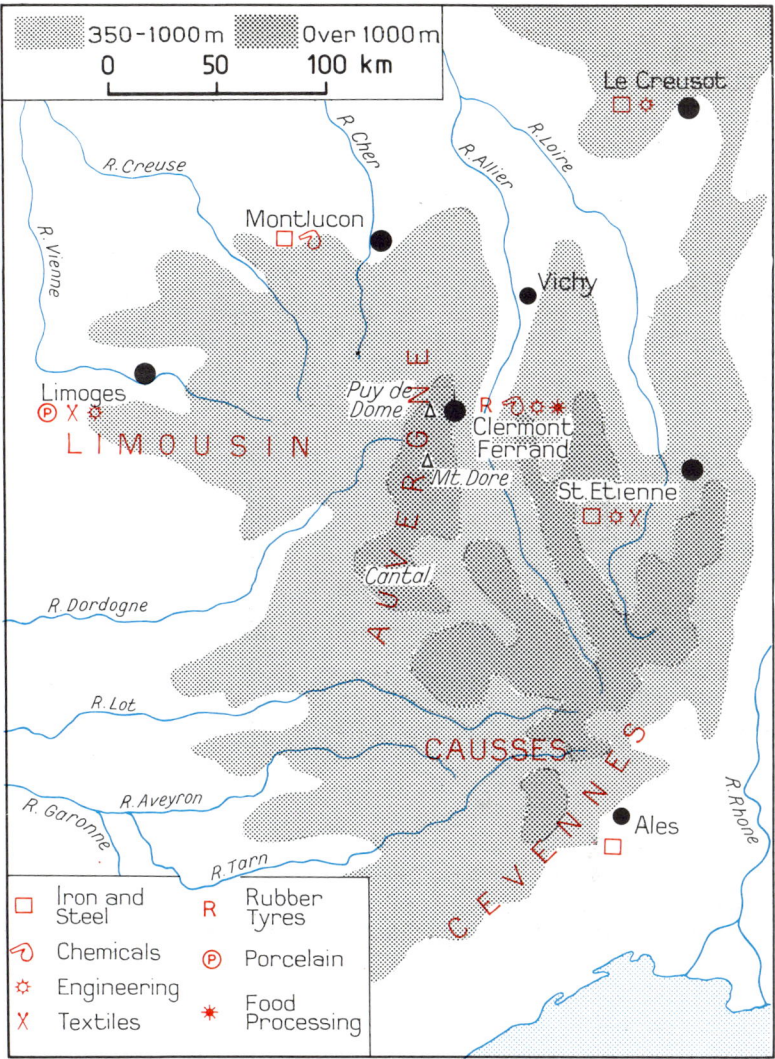

FIG. 8.8 **The Massif Central**

have weathered into fertile soil and provide excellent pasture for cattle and some good arable land, but elsewhere the ground is covered by heath or peat bog. The largest town of the Auvergne is *Clermont Ferrand*, an old-established market town and the home of the giant Michelin tyre company, which started here as a family concern in the nineteenth century. It also has food processing, chemical and engineering industries so that the urban area has a prosperous economy.

The eastern and south-eastern mountain belt contains some of the highest and most rugged country including the *Cevennes*, which end in a fault scarp to the south-east. Much of the land is infertile and used only for cattle rearing, but there are a number of down-faulted depressions in which there is mixed farming, with cattle, grain, root crops and even vineyards. In the south, where the summers are warmer and drier due to Mediterranean influences, sheep and goats are more numerous, and olives and grape vines are grown on terraced slopes. Some of the depressions contain small coalfields and these

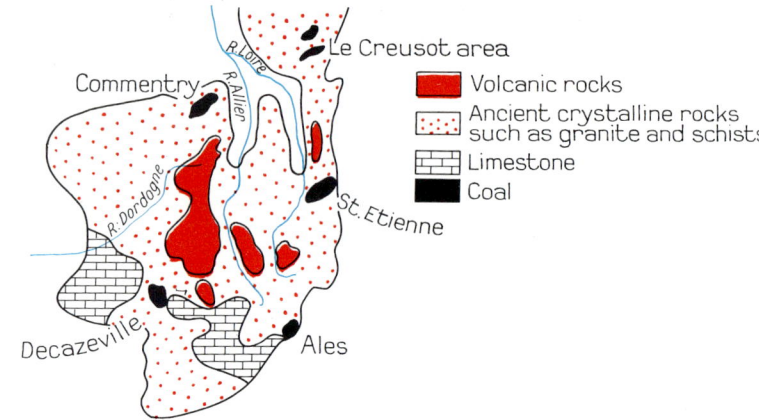

FIG. 8.9 **Simplified geology of the Massif Central**

◄ **The Causses**    *Photo Société Roquefort*

66

**Landscape near Le Puy**

*Photo ESSO*

Describe and account for the physical features in these photographs, and explain the relationship between the types of rock and the land use.

have led to the growth of industries as at *St Etienne*, which was originally concerned with silk and other textiles. Today, besides the iron and steel plant, there are new and expanding industrial estates sited away from the town.

The last sub-region consists of the basins of the Allier and upper Loire. These are rift valleys on whose floors are alluvial deposits derived from volcanic rocks, giving fertile soils on which there are fields of wheat and sugar beet as well as orchards, vineyards and cattle pastures.

In general, the Massif Central is thinly populated apart from a few large towns. Rural employment is largely dependent on livestock farming and on limited areas of arable land. There is also employment in connection with the tourist industry in the scenic areas and in the spas, which include *Vichy*, whose mineral waters are bottled and sold all over France. The availability of hydro-electricity in the valleys of the Dordogne and Lot, where a number of large dams have been constructed, has led to some industrial development, but on the whole the Massif Central is a region of depopulation, with large

67

**Legend:**
- Over 1000 m
- 250–1000 m
- ☐ Steel
- P Oil Refining
- ★ Food Processing
- T Woodworking Industries
- ⛴ Shipbuilding
- ⌒ Chemicals
- ✦ Engineering
- ✄ Textiles
- ⊗ Holiday Resorts
- Main Irrigated Areas

0    100 km

**Map labels:** Dijon, Besançon, R. Doubs, R. Saône, Côte d'Or, Jura, Lake Geneva, Mâcon, Pays de Dombes, Genissiat Dam, Chamonix, Δ Mt. Blanc, Lyons, R. Rhône, High Alps, Mt. Cenis Pass, Tournon, R. Isère, Grenoble, Valence, Montélimar, Donzère, R. Aveyron, R. Tarn, Avignon, R. Durance, Nimes, Nice, Monte Carlo, Arles, Provence, Cannes, Montpellier, Crau, Camargue, Fos, BAUXITE, Marseilles, Riviera, Languedoc, Bèziers, Sète, Narbonne, Toulon, Coastal Languedoc Holiday Resorts, Perpignan

numbers of young people emigrating to find work in the richer parts of France. Depopulation has been most marked in the west and the fact that the remaining farmers have only a small proportion of young men in their number will cause serious problems in the future.

## The Rhône-Saône Valley

The Rhône-Saône valley is a fault-bounded depression which provides an important line of communication between the Mediterranean Sea and northern Europe. The valley is followed by motorway, rail and pipeline. The Saône provides canal links from the Rhône to the Rhine, Seine, Marne and Loire. In recent years the traditionally turbulent and fast flowing Rhône, by a system of locks and by-pass canals, has been transformed into a major waterway.

The Saône valley is a broad trough about 65 km wide, much of it occupied by cattle pasture and arable land on which wheat, maize, sugar beet and potatoes are grown. The southern part is an infertile area of marsh and moor known as the *Pays de Dombes*, which has developed on pebbly glacial material, but even here some parts are being reclaimed for farming, including a large market gardening area near Lyons.

Overlooking the Saône from the west is the *Côte d'Or*, a limestone scarp along which is a narrow belt of terraced vineyards where the high quality Burgundy wines are produced. Near the foot of the scarp is the town of *Dijon* which lies where routes from southern France, Switzerland and the Rhine converge on their way to Paris. It is the headquarters of the local wine industry, and its other manufactures are foodstuffs (including the well-known Dijon mustard), precision engineering, tobacco and footwear.

**Genissiat Barrage** ►

◄ FIG. 8.10

*R. Henrard*

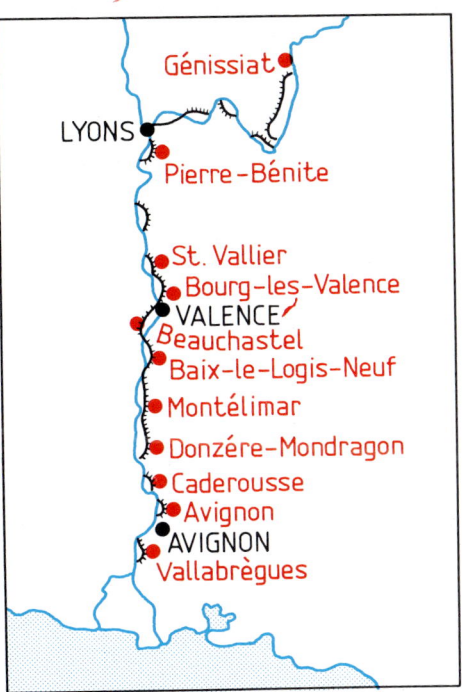

FIG. 8.11 **Hydro-electric power stations and diversion canals on the River Rhône**

The Rhône valley below Lyons is in distinct contrast with the Saône. The river has cut down into an uplifted area and occupies a number of small basins alternating with narrow gorges. Besides improving navigation, the dams provide sites for the generation of electricity and ensure a regular supply of water for irrigation.

To the south of Donzere the valley widens out and the influence of the Mediterranean climate makes itself felt, and hot, dry summers are typical. The enormous increase in land under irrigation is changing

**Pierre Benite (Rhône below Lyons)**

*Photographie EDF*

Describe this scene and that on the previous page, then explain the differences in the construction of the two hydro-electric stations. What part has hydro-electricity played in the development of French industry since the war?

the appearance of large tracts of land. The maquis-type vegetation survives in some areas but land producing almonds, peaches and, especially, a great variety of vegetables, mainly for the Paris market, dominates. An unpleasant aspect of the climate is the cold Mistral, a wind which frequently blows in winter and spring down the Rhône valley towards depressions in the Mediterranean. The wind is a menace to tender crops and these are generally protected by thick cypress hedges and bamboo fences.

New chemical and metallurgical industries, based on the abundant supplies of hydro-electricity, have developed along the Rhône valley, where there are a number of historic towns including the ancient bridge town of *Avignon*.

*Lyons* (1,075.000), the second city of France, lies at the confluence of the Rhône and Saône and is a major route centre. It had an early silk industry which is still carried on although most of the raw silk is now imported from Japan and Italy. The silk industry has been overtaken by the synthetic fibre industry, and there are many other industries including chemical works and a Citroen car works. Power for these industries is derived from the Genissiat hydro-electric station. Lyons is also a river port and can be reached by vessels of 1,500 tonnes as a result of recent improvements to the Rhône waterway.

## The Mediterranean Coastlands

The River Rhône has brought down vast quantities of material, much of which has been deposited at its mouth to form a delta. Between the two main distributaries are the lagoons and salt marshes of the *Camargue*. In the past this has been a desolate area with some sheep, cattle and horses, but in recent years parts have been reclaimed for vineyards, market gardens and rice cultivation. East of the Camargue is the *Crau* which consists of pebbles and boulders brought down by a former course of the Durance. Much land is used as winter grazing for sheep, some of which are taken to the Alps for the summer. Here also, however, there has been reclamation based upon irrigation schemes, and fruit and vegetable cultivation is becoming increasingly significant.

To the east of the delta and well clear of the silt is *Marseilles* (970,000), France's third city and first port. It lies on a sheltered bay and is linked to the Rhône by a canal which passes through a tunnel and then across a large inlet called the Etang de Berre. Marseilles grew to importance because of its trade with the former French possessions in Africa and the Far East (via the Suez Canal). Like most large ports, it has ship repairing and marine engineering industries, and other industries concerned with the processing of imported food-stuffs and raw materials including sugar refining, flour milling, and the manufacture of soap, margarine, paper and chemicals. The Etang de Berre and the Gulf of Fos have become major sites for oil storage and refining. Many chemical plants have been constructed in the area, relying on the oil refineries for raw material. In addition the iron and steel plant at Fos has also assisted in making this a major growth area in France.

East of Marseilles is the naval base of *Toulon*, with a deep, land-locked harbour, and from here as far as the Italian frontier the coastal belt is known as the *Riviera*. This large holiday area has the advantages of a mild, sunny climate, a southern aspect with mountains to protect it from the Mistral, and numerous rocky headlands and sheltered bays. It is easily accessible from the densely populated regions of North-West Europe via the Rhône valley and has numerous resorts including *Nice*, *Cannes* and *Monte Carlo* (which lies in the small principality of Monaco). Farming is also important along the coast, especially the intensive cultivation of early vegetables, salad crops, fruit and flowers.

Inland are the hills and mountains of Provence including the Maritime Alps in the east. Both cattle and sheep graze the hill slopes, and in the valleys are orchards, vineyards and irrigated holdings specialising in the production of flowers.

To the west of the Rhône delta lies the area known as *Languedoc*. The coast here is low-lying and lined with sand dunes, behind which are lagoons and salt marshes. In many places marshes have been drained to provide fertile agricultural land, and in some places salt is obtained by the evaporation of sea water. The only port here is *Sète*, which has oil refining and shipbuilding industries.

The lowlands of Languedoc are France's most important wine-producing region. The growers concentrate on quantity rather than quality and most produce a coarse, red wine known as 'vin ordinaire'. Some areas grow only vines, and others have vineyards along with wheat and olives. However in recent years there has been a great increase in market gardening based on irrigation. To the north the land rises to limestone hills, covered with a low scrub, which provide only poor land for farming. The old market town of *Montpellier* lies near the junction of the hills and plain.

There are Government development projects designed to create new seaside resorts to make the Languedoc coast a major tourist area.

**Irrigated orchards near the Rhône Delta** *French Government Tourist Office*
The view is from the southern foothills of the Alps, looking towards the plain of the Crau.

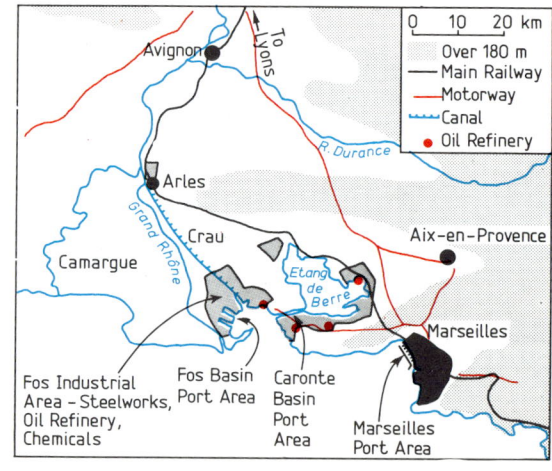

FIG. 8.12 **Marseilles and the Rhône estuary area**

**Marseilles** *Aerofilms* ▶

# CORSICA

Corsica is a mountainous Mediterranean island, and is part of the French Republic. The mountains rise to over 2,000 metres in the west, where they end in a rugged coastline, but to the east the land falls away to a low coast bordered by sand dunes and lagoons.

The Mediterranean climate gives rise to a typical maquis-type vegetation of drought-resisting shrubs with occasional forests of evergreen oak and Corsican pine, and groves of olives and chestnuts. There is rough pasture for sheep and goats, but cultivation is restricted to small level tracts in some of the valleys, and to hillside terraces. The island has few natural resources and there is little industrial development, although the magnificent scenery and pleasant climate attract a growing number of tourists. Of the total population of 220,000 over one quarter live in *Ajaccio*, the capital, and *Bastia*, the chief port.

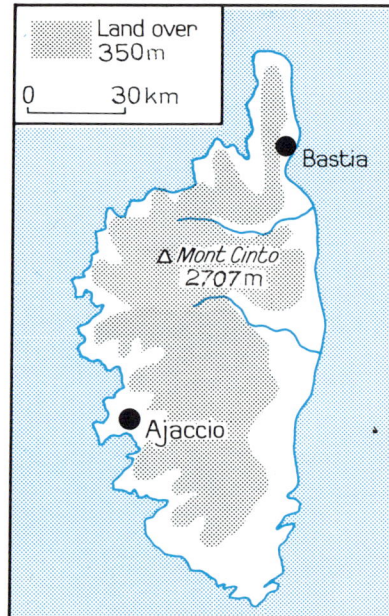

FIG. 8.13 **Corsica**

## The Jura

The Jura Mountains extend from the southern end of the Rhine rift valley to the Rhône valley near Lyons, and lie partly in France and partly in Switzerland. They are built mainly of limestone and in the

**Ste Maxime (French Riviera)**     *French Government Tourist Office*
In what ways does the Mediterranean region of France differ from the rest of France? These photographs may suggest some of the differences.

**Corsican Maquis**     *French Government Tourist Office*

east consist of a series of tight but simple folds, rising to over 1,500 metres above sea level, whilst in the west they become a faulted plateau. Much of the original woodland remains although where it has been cleared, especially in the valleys, dairying and cattle rearing are carried on, some of the milk being made into Gruyère cheese. Forestry provides some employment, and there has been considerable industrial development, particularly at *Besançon* where motor car and rayon factories form a contrast with the traditional domestic crafts such as furniture and clock and watch-making which are now factory industries.

## The Alps

The French Alps consist of the crystalline High Alps in the east including the Italian frontier region, and a series of outer limestone folds called the Pre-Alps which extend to the River Rhône in the west. The High Alps form some massive mountain blocks of which Mont Blanc (4,810 metres) is the highest, with many glaciers including the Mer de Glace, and extensive snowfields which provide excellent winter sports facilities. *Chamonix*, close to Mont Blanc, is one of the most popular resorts and many motorists now use the Mont Blanc tunnel on their way to Italy.

The northern French Alps have a continental climate and are characterised by coniferous forests, fast-flowing streams and rich valley meadows, in contrast to the southern French Alps whose Mediterranean climate results in poor pasture and scrub, with chestnut and olive trees in the valleys, and dried-up river beds during the summer. The types of farming differ also; in the north cattle are kept and fodder crops are grown in the valleys, whilst in the south there are far more sheep than cattle and cultivation in the valleys is mainly restricted to terraces and small plots where cereals, vegetables and fruit are grown with the aid of irrigation. Throughout the Alps the system of transhumance is practised whereby livestock graze the high mountain pastures during the summer, and spend the winter in the valleys.

There has been a remarkable industrial expansion in the Alpine region following the recent construction of large numbers of hydro-electric stations. There are electro-metallurgical and electro-chemical industries in many of the valleys, including the making of steel-alloys using electrical furnaces and the refining of aluminium based on bauxite mined in southern Provence. Old-established industries like forestry and timber-working, paper, textiles and leather-working have retained their importance.

The main industrial centre is *Grenoble*, which lies on the Isère, on the Mt Cenis Pass route into Italy. It is famous for the manufacture of gloves (using both local and imported skins), and also has silk, rayon, cement and metallurgical industries.

## Alsace-Lorraine

After the Franco-Prussian War of 1870 these provinces became part of Germany, but were restored to France by the peace treaty following the First World War.

**The Southern French Alps**    *French Government Tourist Office*
This procession of sheep and goats is making its way to the high pastures for summer grazing. The grass here is too poor for cattle.

*Lorraine* lies between the eastern rim of the Paris Basin and the uplands of the Vosges. It consists of alternate layers of hard and soft rock, with the strata dipping towards the west. The region is drained by the rivers Meuse and Moselle which flow north along clay vales, overlooked by the east-facing limestone scarps of the Côte de Meuse and Côte de Moselle. Farther east the land rises to sandstone hills bordering the Vosges. The uplands have thin, infertile soils and are well-wooded, but the lowlands are moderately fertile when drained.

Lorraine's greatest wealth lies in its minerals. Most important are the huge deposits of iron ore which outcrop at the base of the Côte de Moselle. Although the ore occurs in thick beds and is easily worked, the average metal content of 33 per cent is low. It was neglected until the 1880's because of its high phosphoric content, but then the invention of the Thomas and Gilchrist basic process led to a rapid increase in production. Present annual output is over 40 million tonnes gross (about 12 million tonnes metal content) and some is exported to Belgium, Luxembourg and the Saar, but production is now falling due to competition from foreign high grade ores. A major iron and steel industry, using the local ore, has developed in the region, the principal centres being *Longwy*, *Thionville* and *Nancy*. There is a small coalfield in the north, a continuation of the Saar coalfield, and although it does not produce good coking coal, new techniques have been developed to enable some of it to be made into coke for use in the nearby iron and steel industry. The steel industry in Lorraine has experienced problems in 1977–79 and this, with the contraction of the textile industry has led to the introduction of new industries such as motor-cars and chemicals around Metz and Longwy.

*Nancy* is the largest town in the region, lying where an important north-south route crosses the Strasbourg-Paris route. In addition to its iron and steel industry, it has a chemical industry, making use of local deposits of salt, and a textile industry.

The *Vosges* are a typical horst and rise gradually from Lorraine to rounded granite summits of over 1,000 m, and then fall steeply to

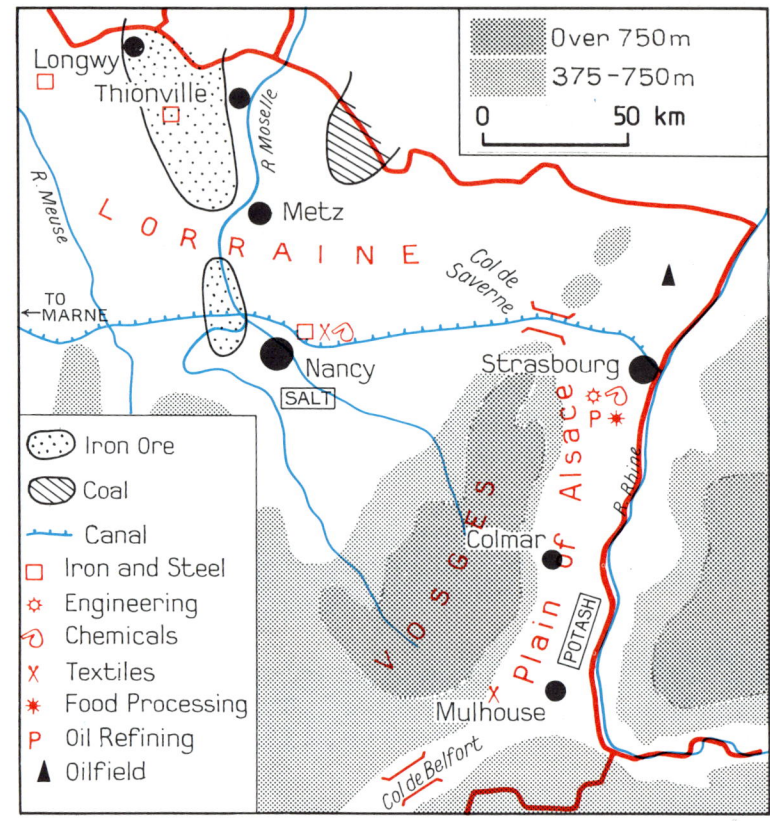

FIG. 8.15 **Alsace and Lorraine**

the Rhine valley. The granite core has a sandstone surround which has been deeply dissected by streams. The steeper slopes are forested

FIG. 8.14 **Simplified section across Alsace and Lorraine**

and the summits provide summer pasture for dairy cattle. The lower slopes are partly cultivated, with cereals, tobacco and vineyards.

Forestry, woodworking, and pulp and paper mills also provide employment, and there is a long tradition of textile manufacturing in a number of small towns.

The *Plain of Alsace* is bordered to the east by the Rhine and to the west by the Vosges, and is the French portion of the Rhine rift valley. Much has been done to improve the navigation of the Rhine, including the construction of short canals by-passing difficult stretches. Three such canals have so far been completed, each with a hydro-electric station.

Alsace is an important region agriculturally. On the lower land beside the Rhine is a waterlogged flood plain which is used as permanent pasture for dairy cattle. Above this are patches of infertile glacial gravels, and higher still are fertile loess-covered terraces which produce excellent crops of vegetables, wheat, sugar beet, hops and tobacco. In the foothills of the Vosges are orchards and vineyards especially where there is a southerly or easterly aspect.

There are large potash deposits between *Mulhouse* and *Colmar* which have given rise to a fertiliser industry whose output has been greatly stepped up in recent years, large quantities being sent down the Rhine to the Netherlands and Belgium. There is also a small oilfield at Pechelbronn. The main industry of Alsace is the manufacture of textiles especially at Mulhouse where besides cotton goods a wide range of other materials is produced.

*Strasbourg* (250,000) is the capital of Alsace, and was a fortress in Roman times. It lies on the River Ill, close to the Rhine, and at the junction of the route along the Rhine and the route westwards to Paris via the Col de Saverne. It is a major river and canal port, and has many industries including oil refining, chemicals, engineering, food processing and brewing. It is also the seat of the European Parliament.

## The Nord Coalfield

From the Meuse valley coalfield in Belgium the Coal Measures extend into France in a narrow band some 10 to 16 kilometres wide round *Valenciennes* and *Lens*. The coalfield has declined dramatically and now produces only 7 million tonnes per year, a third of France's coal. Many of the smaller pits have been closed and the others have been modernised since the war, with automation playing a big part in the mining. Much of the coal is used locally in thermal power stations and coke ovens, but some is sent by waterway to the Paris region and to Lorraine.

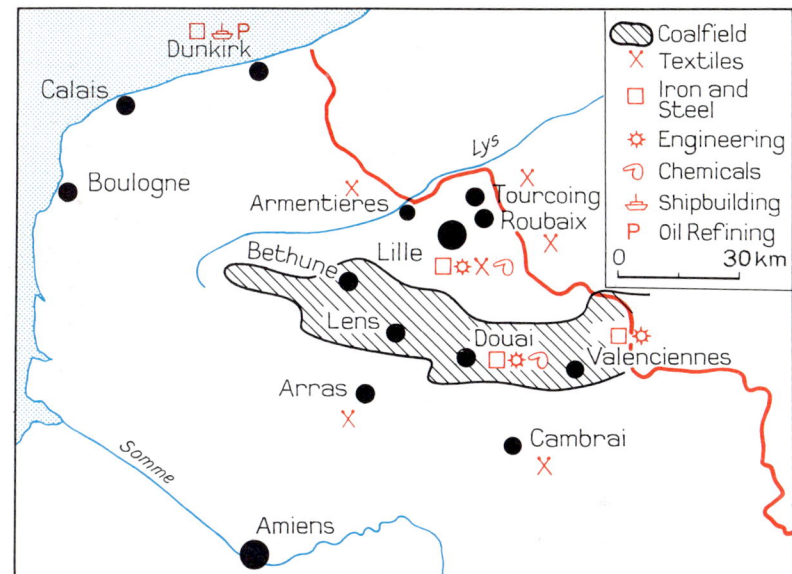

FIG. 8.16 **The Industrial North of France**

There was an early textile industry in the area, using wool from local sheep, and flax. With the Industrial Revolution the use of coal as a source of power made this the major textile region of France. Cotton became the leading textile but recently a great variety of synthetic fibres have been introduced. Lille, for long the centre of the textile industry, has had major industrial developments in recent years with the expansion of car manufacturing most significant. The town at the same time has become an attractive commercial centre.

There is also a large iron and steel industry, based on local coal and Lorraine iron, extending from Valenciennes to Douai, and associated with it are many branches of the engineering industry. Lille, the main town of the region, is the centre of the textile industry and also has engineering and chemical works.

Industry has recently expanded to the coast, especially at *Dunkirk*, which is the chief port of the region and has large steelworks (using imported ore), as well as shipbuilding and oil refining. The port has new terminals to cope with oil tankers and iron ore carriers. *Calais* and *Boulogne* are Channel ferry ports.

77

## The Paris Basin

The Paris Basin is the largest and most important region in France. It has an undulating relief of low plateaus and plains, and is drained by the Seine and middle Loire. Structurally it is a broad, shallow syncline consisting of layers of sedimentary rock, one inside the other. In the centre are Tertiary sandstones and limestones, which are surrounded by successive belts of chalk, clay and Jurassic limestone, the edges of the chalk and limestone standing out most prominently in the east and south-east.

It has a maritime climate, with moderate rainfall and this, combined with the variety of soils, has made the Paris Basin the most productive farming area in France. Nature has provided easy route-ways too, and Paris has become the focus of the region, and of all France. It is convenient to divide the Paris Basin into a number of sub-regions.

### The Middle Loire Region

Most of this region is infertile plateau country, with some farming on improved land but also large stretches of moor and woodland. In contrast, the Loire valley itself is generally fertile and has a prosperous agriculture. The river alluvium provides pasture land for cattle and is bordered by gently-sloping terraces which are well-cultivated with wheat, potatoes, vegetables and orchards. There are many historic towns and chateaux which have led to a considerable tourist industry. *Orleans* lies where the right-angled bend of the Loire brings it nearest to Paris, and is a route centre and market town, with agricultural engineering and food processing industries. *Tours* is another market town with similar industries, and lies where the Paris Bordeaux route crosses

**Factories in Lille**

*Aerofilms*

Compare this scene with that of the Citroen factory on page 62, pointing out the main factors responsible for the location of the industries in both cases.

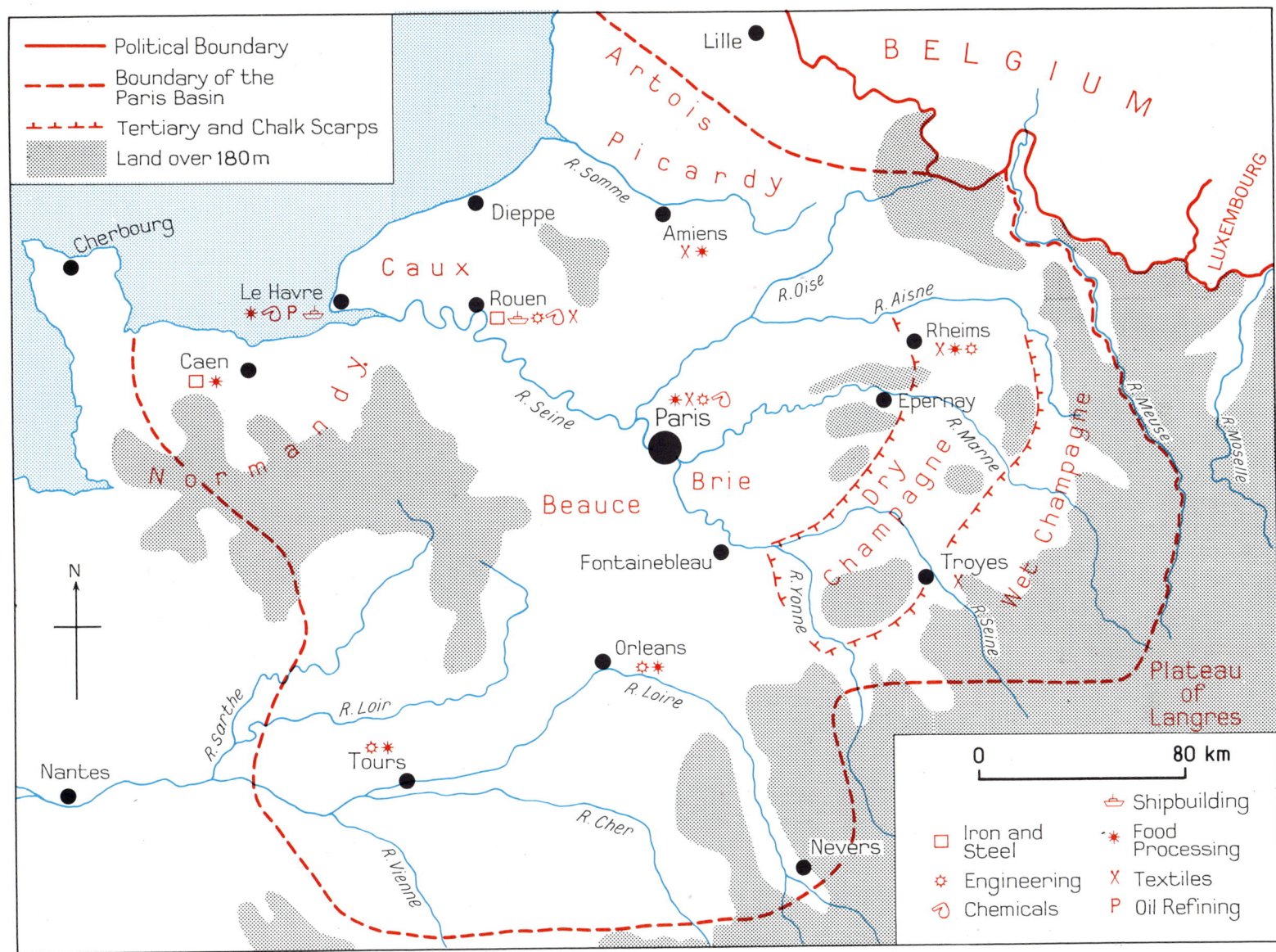

FIG. 8.17 **The Paris Basin**

the Loire. Both Orleans and Tours are experiencing industrial expansion after a long period of stagnation.

## The Eastern Scarplands

In the eastern part of the Paris Basin the erosion of the alternating layers of hard and soft rocks has produced a landscape of scarps, dip slopes and vales. The Falaise de l'Ile de France marks the eastern edge of the Tertiary limestones, and overlooks a dry, treeless chalk plateau known as the dry Champagne. For many centuries the poor, chalky soils were used for sheep grazing, but today, with the use of fertilisers, there is large-scale cultivation of cereals, especially barley. Settlements are concentrated mainly in the valleys which cut across the region, and here alluvial soils produce good grassland for cattle.

The famous sparkling wine known as Champagne comes from the scarp slope of the Falaise and from the south-east facing sides of valleys leading off from the scarp. The area is near the northern limit of wine production, but it benefits from a warm soil and sunny aspect, though even more important than these is the great skill of the wine makers. *Rheims* and *Epernay* are the organising centres of the industry. Rheims is a gap town on an important route to Paris, and has textile, food processing and engineering industries.

The chalk ends in a prominent scarp and is succeeded by a broad clay vale called the wet Champagne. In the past much of this has been woodland or marsh, but large areas have now been cleared and drained, providing good pastures for beef and dairy cattle. Beyond the wet Champagne are Jurassic limestone uplands ending to the east in the Côte de Meuse on the borders of Lorraine, and to the south-east in the Plateau of Langres.

## The Northern Plateaus

The chalk plateau continues north-west from the dry Champagne through Picardy and Artois and into the Caux region of upper Normandy. Here the land is much more fertile than the dry Champagne since extensive deposits of loess overlie the chalk. There is large-scale arable farming, with wheat and sugar beet as the main crops, and also flax, hops and fodder crops. Cattle are more important than sheep and are especially numerous in the broad valleys which are mainly under permanent pasture. One of the chief towns here is *Amiens*, which lies at a crossing point of the Somme and has a long tradition of textile manufacture although today it also has important food processing and footwear industries.

**Champagne**     *French Government Tourist Office*
These vineyards are among the most northerly in Europe, but they are planted so as to obtain a maximum of sunlight, and produce one of the most famous wines in the world.

**Countryside near Boulogne**     *French Government Tourist Office*
This is a rolling plateau of chalk, covered by a layer of fertile loess. Large-scale arable farming is practised, with wheat and sugar beet as the chief crops.

**Rouen**
Compare Rouen with
Marseilles (page 73)
from the points of
view of site, situation
and trade. Illustrate
your answer by means
of sketch maps.

*Aerofilms*

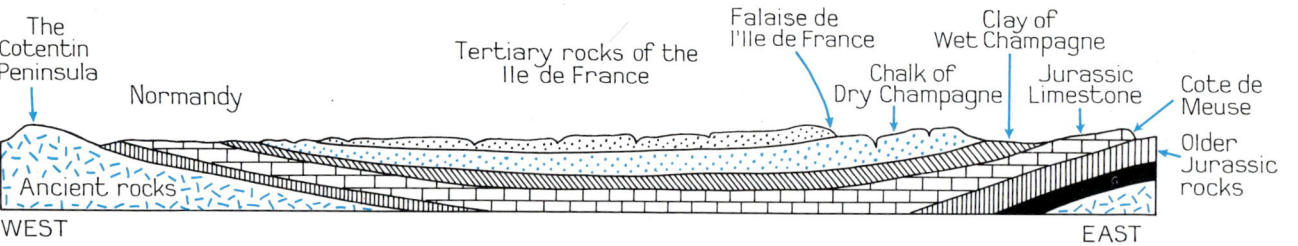

FIG. 8.18 **Simplified section across the Paris Basin**

The Cotentin Peninsula

Normandy

Tertiary rocks of the Ile de France

Falaise de l'Ile de France

Chalk of Dry Champagne

Clay of Wet Champagne

Jurassic Limestone

Cote de Meuse

Older Jurassic rocks

Ancient rocks

WEST

EAST

South of the broad flood plain of the Seine, lower Normandy forms a contrast with the rest of the Paris Basin, since the chalk here is partly covered by superficial deposits of clay-with-flints. As a result, there is a considerable proportion of grassland and cattle are kept both for beef and milk. Much of the milk is made into butter and cheese, including the well-known Camembert. Apples are grown for cider-making. The main town is *Caen* which has an iron and steel industry using local iron ore deposits.

*Le Havre* lies on the northern side of the Seine estuary and is the second port of France and main outlet for the Paris Basin. It is used by passenger liners and has a ferry service to Southampton, as well as a large cargo trade. Mineral oil is its principal import and there are oil-refineries nearby. Other industries are food processing, chemical manufacture and shipbuilding. Vast new industrial estates are making Le Havre a major manufacturing centre. *Rouen* is 60 kilometres from the sea and was formerly a more important port than Le Havre. It is still a leading cargo port although large vessels can reach it only at high tide. It imports raw materials for its iron and steel, cotton textile, shipbuilding and heavy chemical industries, and also for the industries of the Paris region. Heavy, bulky goods such as coal are transhipped into Seine barges for onward despatch.

### The Ile de France

The central part of the Paris Basin is known as the Ile de France and consists of young Tertiary rocks, mainly sandstones and limestones, which form low plateaus in which the rivers have cut deeply-entrenched valleys. In general the plateaus are covered with loess and are devoted to arable farming, particularly wheat. There is little pasture land and most of the cattle are kept in the farmyard. Sheep, which were numerous, have largely disappeared so that farmers can

concentrate on cereal production. On the alluvial soils, especially on the outskirts of Paris, there is much market gardening.

The Ile de France contains a number of distinctive 'pays' (regions) including Beauce and Brie. Beauce is a flat limestone plateau to the south-west of Paris, with a loess cover which is fertile in spite of the absence of streams. It has been called the 'granary of France', and has large farms with crops of wheat, maize, barley and sugar beet. Brie, to the south-east of Paris, has a clay capping over the limestone, giving rich grasslands which are mainly used for dairy farming. Quite a different type of landscape occurs in the sandy areas where considerable stretches of woodland remain, as in the forest of Fontainebleau, which serves as a playground for the people of Paris.

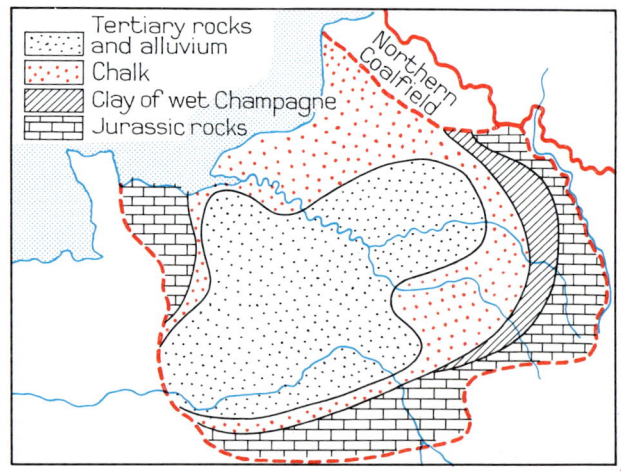

Tertiary rocks and alluvium

Chalk

Clay of wet Champagne

Jurassic rocks

Northern Coalfield

FIG. 8.19 **Rocks of the Paris Basin**

## Paris

Paris first grew up on a small island in the Seine which provided an easily-defended site and a good bridging point. The island is now called the Ile de la Cité, and lies down-river from the Seine-Marne confluence. The city's later development owes much to its nodal position within the Paris Basin. The principal rivers, including the Seine, Oise, Marne and Yonne, have cut gaps in the surrounding hills and are followed by roads and railways which converge on Paris. It was from the first a flourishing river port, and there is still much barge traffic bringing in bulk cargoes of coal, oil, foodstuffs and building materials from Rouen and other parts of France.

The city itself has a population of 3 million, with 8·8 million in Greater Paris. In addition to being capital of France, it is the outstanding commercial and cultural centre of the country, and a great manufacturing city specialising in high quality articles which require skilled workmanship, for example clothing, furniture and jewellery, along with a wide range of food and drink products. In addition, modern engineering and chemical industries have developed both in the city and in the suburbs, including large Renault and Citroen car works.

Paris is an attractive city with wide, tree-lined boulevards, many famous buildings, and a wealth of entertainment which have made it one of the world's greatest tourist centres.

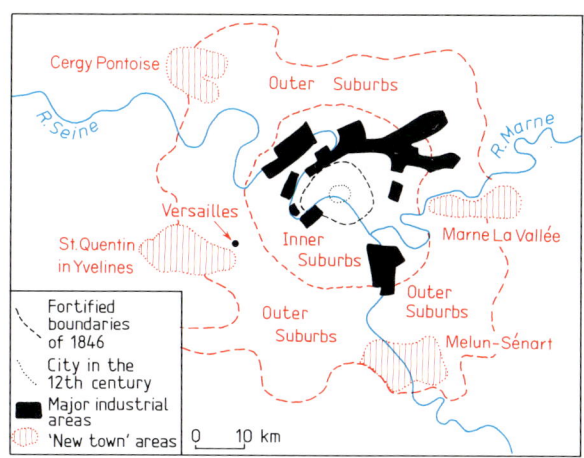

FIG. 8.21 **The site and development of Paris**

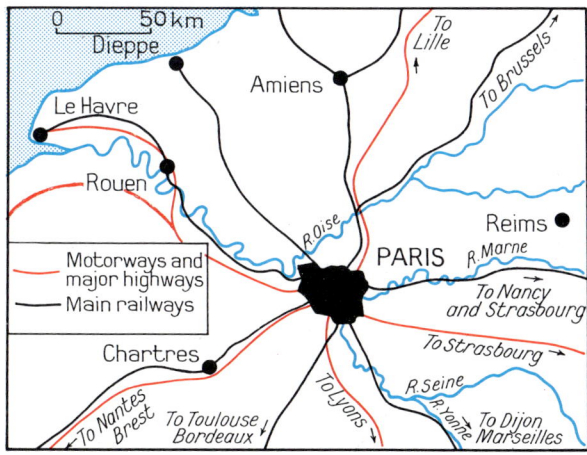

FIG. 8.20 **The position of Paris**

## Exercises

*Answer in note form:*

1. Account for the differences in the climatic conditions in Brittany, the French Riviera and the Rhine rift valley.

2. Locate and describe the scenery associated with the Landes, Camargue, Bocage and Maquis.

3. Name an area in France important for the production of each of the following: wheat, maize, cheese and fish. For each area state how the geographical conditions are suitable for the product.

4. Using sketch maps, with annotations, explain the importance of the Col de Poitou, Col de Carcassonne and Col de Belfort.

*Essay Questions*

1. To what extent have power supplies influenced the location of manufacturing industries in France?

2. Explain the relationship between soil and agriculture in the Paris Basin.

3. Write an account of vine-growing and wine production in France, illustrated by a sketch map to show the main wine-producing areas.

4. Describe a journey by river and canal from Le Havre to Marseilles via Dijon.

5. How is the scenery of the Massif Central influenced by the types of rock?

# INTRODUCTION TO GERMANY

After the 1939 to 1945 War, Germany suffered the loss of large territories in the east to Poland and the Soviet Union, and the German-Polish boundary now follows the rivers Oder and Neisse. At first the country was divided into four occupation zones controlled by Great Britain, France, the United States and the U.S.S.R. In 1949 the three western powers merged their zones and the *German Federal Republic* came into being. In the same year a communist system of government was established in the Soviet-occupied zone of East Germany, and in 1955 the U.S.S.R. recognised this as an independent state known as the *German Democratic Republic*. We must therefore treat East and West Germany as two distinct states. Berlin, the former capital of Germany, lies within the boundaries of East Germany, but is itself divided into eastern and western sectors. The eastern sector is now part of East Germany, but the western sector has its own administration, and has close links with the West German authorities.

# 9: THE GERMAN FEDERAL REPUBLIC (West Germany)

West Germany has an area of 250,000 square km and a population of 62 million, making it one of the most densely populated countries in Europe, with an average density of nearly 250 to the square km. Its population has been swollen by large numbers of refugees from former German territories now incorporated in Poland and the U.S.S.R., from Czechoslovakia, and some three million who have fled from East Germany.

The revival of West Germany after the Second World War and its re-emergence as a major industrial nation is remarkable. The period from 1945 to 1950 was one of reconstruction, and since 1950 there has been a period of great expansion so that today West Germany is one of the world's most important industrial nations, and a leading member of the European Economic Community.

## Structure and Relief

The northern part of West Germany belongs to the North European Plain. During the Ice Age the Baltic ice sheet advanced and retreated several times across the region, and has left widespread glacial deposits including morainic gravels and boulder clay. Melt water streams flowing from the edge of the ice sheet also deposited much sandy outwash material, and wind-blown dust settled along the southern margins of the plain to form extensive sheets of loess.

To the south the land rises to a series of uplands including the Middle Rhine Uplands and the Central Uplands, forming a region of complex structure and relief. It was part of the ancient Hercynian fold mountains, and the land was worn down and then greatly faulted as a result of the Alpine movements. Some areas were upraised between faults to produce steep-sided horsts, whilst others subsided to form tectonic depressions, as in the Rhine rift valley. These movements were accompanied by considerable igneous activity.

In the far south are the limestone Bavarian Alps, which are part of the northernmost ranges of the Alpine fold mountains. On their northern flanks is a broad plateau sloping gently down to the Danube valley, known as the Alpine 'foreland'. The foreland is mainly covered by glacial deposits derived from the ice cap which spread out from the Alps during the Ice Age.

## Climate

West Germany's climate is subject to both maritime and continental influences. Winters are mild in the north-west, which has mean January temperatures above freezing point, but become more severe to the east and south where frost and snow may last for many weeks. Summers are fairly cool everywhere, with mean July temperatures between $17°$ and $18°C$, the warmest areas being in some of the more sheltered Rhineland valleys.

The annual precipitation amounts to between 685 and 750 mm on the northern plain, which is fully exposed to the rain-bearing westerly winds, but can be as low as 500 mm in some of the valleys and basins farther south. The uplands have the highest precipitation which generally exceeds 1,000 mm and reaches 2,000 mm in the Bavarian Alps. The precipitation is fairly evenly distributed throughout the year, but there is a slight summer maximum.

## Farming

About 7 per cent of the working population are employed in farming, but because of the large tracts of bleak upland and infertile, sandy lowlands, food production can satisfy only about three-quarters of the country's requirements.

West Germany is a land of small farms whose average size is about 6 hectares. In some areas fragmentation has occurred whereby a farm may consist of many scattered strips of land. Government policy is aimed at the consolidation of such scattered holdings to form more efficient units on which machinery and other modern techniques can be used.

The basis of agriculture is livestock farming. Both pigs and cattle are important throughout the country and most of them are sty or stall-fed. Only the moist coastal lowlands and lower valleys have much grass. Rye and potatoes are the chief crops on the poorer soils of the North German Plain, whilst wheat and sugar beet do well on the richer loess soils. The warmer climate of the Rhineland valleys is particularly favourable for vineyards and orchards. Market gardening and dairying are important activities in the vicinity of most large towns and industrial regions.

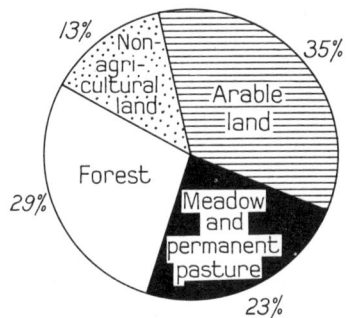

FIG. 9.1 **West Germany—land use**

## Forestry and Fishing

Over a quarter of West Germany is forested, much of it the result of planned afforestation which has been undertaken on a large scale both on the sands and gravels of the North German Plain and in the hill country of the centre and south. The trees are mainly conifers, including fir, spruce and pine, but some oak, beech and other deciduous trees are found on the lower hill slopes in the Rhineland and elsewhere. Forestry is under strict government control and the timber is much used in constructional work, in furniture-making and wood carving, as a raw material in the paper and rayon industries, and as fuel.

The fishing industry is organised along modern lines and trawlers operate in the North Sea, Baltic Sea, and in Icelandic and Arctic waters, from a number of ports including Bremerhaven and Cuxhaven. Herring drifters are based on the smaller Weser ports and Emden.

## Industry

In Germany the Industrial Revolution came later than in Britain, which was an advantage for it was able to make use of the latest scientific and technical knowledge in the development of its resources.

West Germany has large reserves of bituminous coal, with an annual output of some 97 million tonnes, eighty per cent of which comes from the great Ruhr coalfield and most of the rest from the Saar and Aachen fields. In addition, there are valuable deposits of lignite especially near Cologne. Iron ore occurs in the middle Rhine area and round Salzgitter, but much of the richer ores have been worked out and practically all of the country's requirements now have to be imported from Sweden, Brazil, Lorraine and elsewhere. Salt and potash deposits in the southern part of the North German Plain have played a large part in the development of the chemical industry, and there are small workings of lead, zinc and copper in the Central Uplands.

The Federal Republic produces small quantities of mineral oil in the northern plain, but the annual production of 5 million tonnes makes up only a small proportion of her requirements, which are over 100 million tonnes per annum. Imported oil is brought by pipeline from Marseilles, Rotterdam and Wilhelmshaven to refineries in the interior of the country at such places as Karlsruhe and Ingolstadt. There has been a sharp increase in the production of natural gas which is now competing with both coal and oil as a source of power. Most of the gas fields are in north Germany.

Steel production rose rapidly after the war, exceeding 40 million tonnes in recent years. At present West Germany occupies fourth place among the producing countries, after the United States, U.S.S.R. and Japan. Most of the iron and steel and heavy engineering industries are concentrated on the Ruhr coalfield, but there are important centres

FIG. 9.2 **The regions of West Germany**

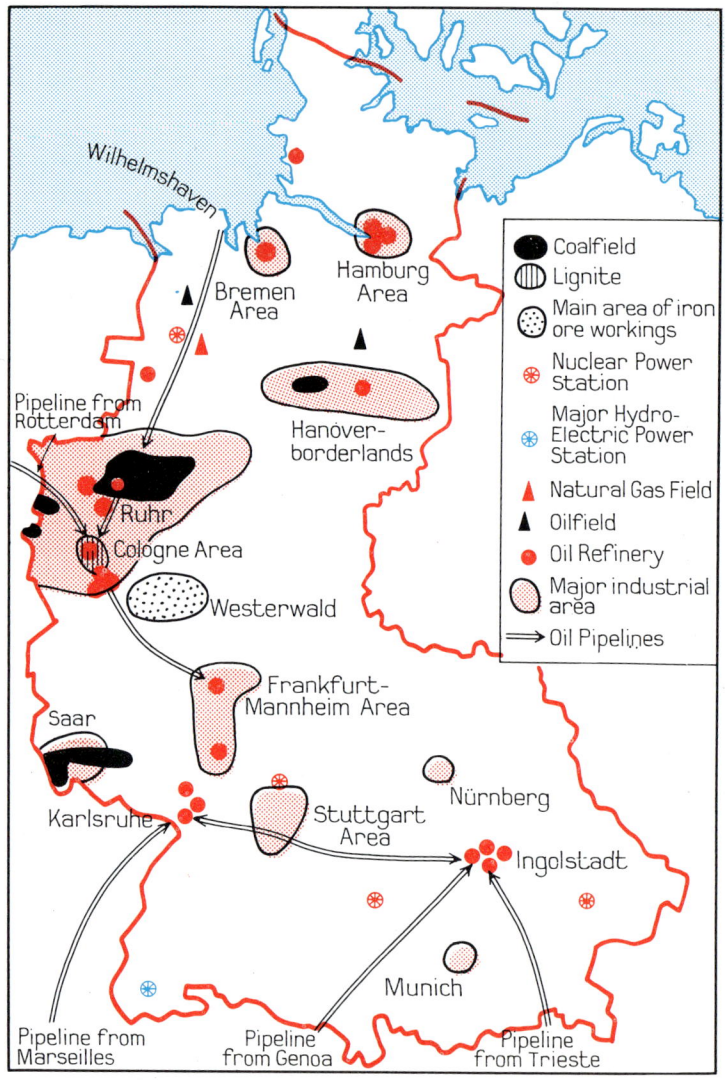

FIG. 9.3 **Major industrial resources of West Germany**

elsewhere including the Saar coalfield and the Brunswick-Salzgitter area. Shipbuilding is carried on at Hamburg, and in the Bremen-Bremerhaven and Kiel areas, whilst light engineering is widely dispersed throughout the country.

The chemical industry has also grown at a fast rate during recent years. Its principal raw materials are salt, potash and coal, and for the petro-chemical industries, mineral oil. Again the Ruhr and the surrounding area is the leading chemical manufacturing region, although Frankfurt and the Mannheim-Heidelberg areas are of considerable importance.

Other industries, including the manufacture of textiles and clothing, food processing and brewing, are carried on both in the main industrial areas and in smaller towns in other parts of the country.

## Foreign Trade

West Germany is one of the great trading nations of the world, and its ability to import the large quantities of food, fuel and raw materials needed to sustain the high standard of living depends on a continued high level of exports. The principal exports consist of engineering products, including motor vehicles, machinery and electrical apparatus, iron and steel, optical and scientific equipment, chemicals, textiles and coal. Much of this trade is carried on with other European countries, especially those belonging to the Common Market, the United States and Canada, and the 'developing' countries of South America, Africa and South-East Asia.

# THE REGIONS OF WEST GERMANY

## The North German Plain

### The Coastlands

The North Sea and Baltic coastlands are low-lying and there are extensive coastal marshes which have been drained and protected by dykes, forming rich pastures for both dairy and beef cattle. The coast is broken by a number of large estuaries including those of the Ems, Weser and Elbe. The sandy Frisian Islands lie only a short distance offshore and provide good sites for holiday resorts.

*Hamburg* (1,750,000) lies about 120 kilometres up the Elbe, and is West Germany's largest city and leading port. Its natural hinterland, the Elbe basin, was lost as a result of the post-war partition of Germany, but in spite of this Hamburg has continued to develop and now handles about half of the country's foreign trade. It is a great industrial centre concerned with shipbuilding, engineering, the processing of imported foodstuffs and varied light industries. *Cuxhaven* has expanded as an outport for Hamburg as well as being a fishing port.

*Bremen* (600,000) lies on the Weser, about 65 kilometres from the sea. It grew up as a bridge town and has become West Germany's second port, with numerous industries including shipbuilding, engineering, textiles, and a large steelworks based on imported ore. The silting up of the estuary has proved a handicap and this has led to the development of *Bremerhaven*, which is West Germany's chief passenger port, and other outports handling vegetable oils, rubber and chemicals which are processed in the area.

*Emden* lies on the estuary of the Ems and is linked to the Ruhr by the Dortmund-Ems Canal. Its imports include Swedish iron ore and timber, and one of its main exports is coal. *Kiel* is the largest town on the Baltic coast and lies at the northern end of the Kiel Ship Canal. It was formerly a great naval base, but is now a commercial port with shipbuilding and light industries.

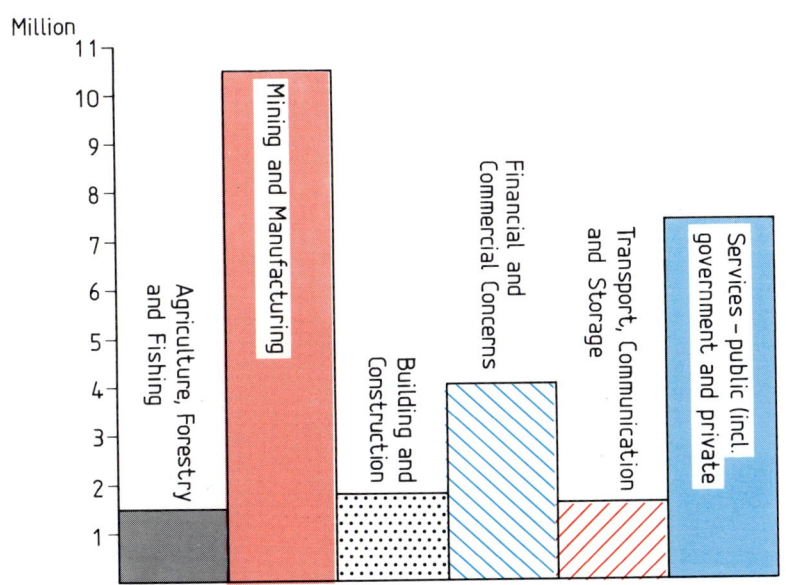

FIG. 9.4 **Main sectors of employment in West Germany**

**Hamburg**

*Inter Nationes, Bonn*

A view of the harbour, with power stations, ship repairing and marine engineering works.

◄ **North German Plain**

*J. Allan Cash*

Looking south from the Baltic coast, between the ports of Kiel and Lubeck (exercise on page 101).

**Volkswagen Works at Wolfsburg**

This is Europe's largest car works. The photograph shows one of the assembly lines. Although many of the processes are fully automated, some of the work must still be done by hand. ►

*Deutsche Presse Agentor, Frankfurt*

### The Geest

Inland the North German Plain consists mainly of low plateaus of outwash sands and gravels known as *geest*. In its natural state the geest is heathland, with a heather and gorse vegetation, and can be used only for sheep grazing. However, considerable areas have been planted with pinewoods, and much of the rest has been reclaimed by deep ploughing, which mixes the sands with the underlying layers of peat, followed by liberal applications of fertilisers. Oats, rye and potatoes are grown, and pigs are kept, on this improved land.

There are numerous surface depressions in the geest containing peat bogs. These peat bogs are known as *moore*, and many of them have been drained by the digging of canals and ditches to provide good cattle pasture. Some of the peat is dried for fuel.

One of the main geest areas is the Oldenburg geest to the west of the Weser valley, and here in the centre of a large reclaimed area is the old town of *Oldenburg*, whose food processing and agricultural machinery industries reflect its links with local farming. An even more extensive area of geest is the Luneburg Heath, between the Weser and Elbe, and a third large area occupies the western part of Schleswig-Holstein. A number of morainic ridges, remnants of the terminal moraine of the Baltic ice sheet, cross the eastern part of Schleswig-Holstein.

Since 1945 great efforts have been made to locate mineral oil deposits in West Germany, and two important oilfields are now being worked on the Northern Plain, one on the southern edge of the Luneburg Heath and the other in the Ems valley.

### The Borde Zone

Along the southern margins of the North German Plain is a wide zone of Börde (border) country. Here large tracts, especially round Hanover and Brunswick, have fertile, loess soils and are under intensive arable farming. Wheat and sugar beet are the main crops, and cattle and sheep are kept as part of the arable farming system, living on roots, hay and beet pulp. To the west lies the semi-circular Munster Bay, a lowland overlooked by high ground to the north-east, east and south. The central part of the Bay has heavy, clay soils and is mainly under pasture, grazed by cattle and sheep, and from here large quantities of milk and meat are sent to the nearby Ruhr industrial region.

The Börde zone is also rich in minerals, including potash, salt and lignite, and in the hills near Salzgitter there are thick beds of low-grade iron ore. As a result, there has been a considerable development of chemical and metallurgical industries.

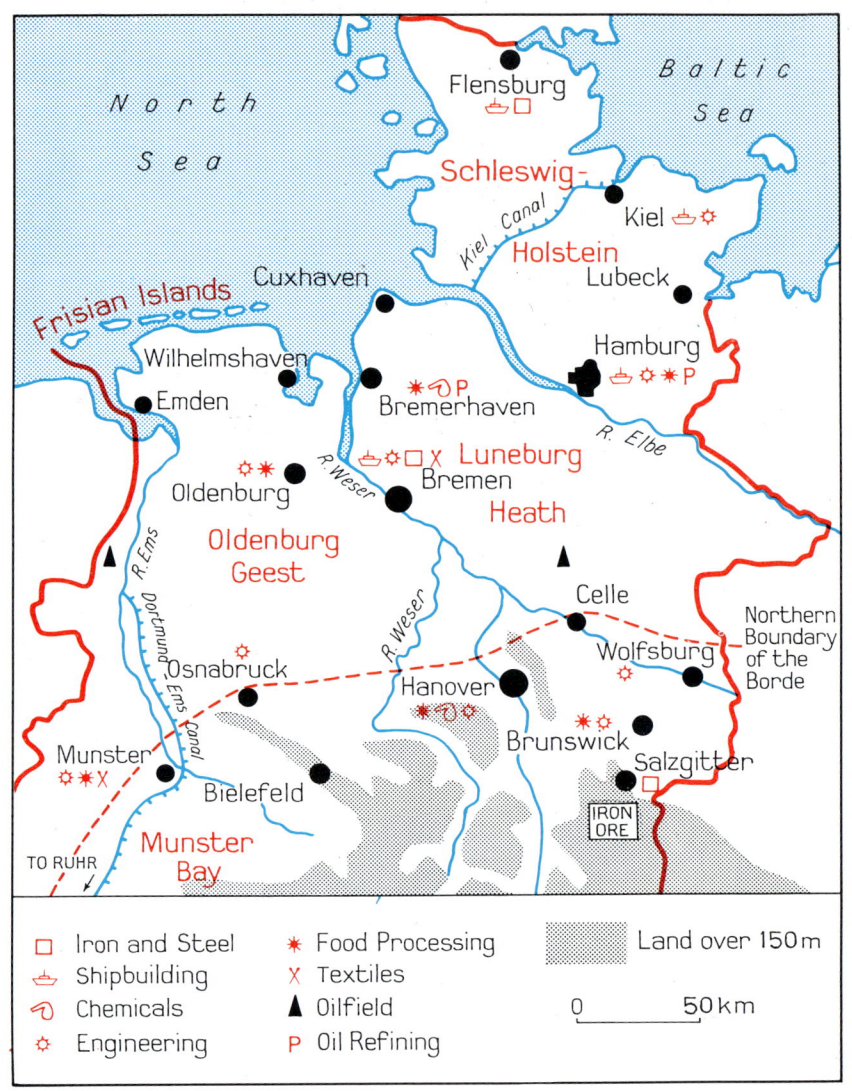

FIG. 9.5 **The North German Plain**

*Hanover* (550,000) is the largest city of the region and has developed as a great market and communications centre, lying where an important route from the hills crosses the east-west route along the Börde zone. Its industries include food processing and the manufacture of fertilisers, rubber and motor vehicles.

*Brunswick* (280,000) is another regional market town with many industries including food processing, precision engineering and motor vehicle manufacture. In contrast to these old towns, there are two relatively new industrial towns, *Salzgitter*, which has a large iron and steel works using local ore, and *Wolfsburg*, with its huge Volkswagen car works.

## The Rhineland

The Rhine valley is the principal north-south routeway of Central Europe, and the river itself is one of the most important waterways in the world. It is navigable from Rheinfelden, upstream from Basle in Switzerland, to the sea, a distance of 800 km. Cologne is the head of navigation for ocean-going ships, whilst barges of 2,000 tonnes can reach Basle. From the river, many important tributaries and canals, some in the process of improvement, provide a link with industrial areas in western and southern Germany, eastern France and Switzerland.

Navigation is not without its difficulties. The gorge section has rapids, and variations in the level of the river, with low water in winter and a danger of flooding in spring and early summer, can also bring problems. Considerable improvements have been effected in the rift valley section where the course of the Rhine has been regulated and a straight channel cut through its meanders.

### The Northern Lowlands

From Bonn the lower Rhine flows across a broad lowland as far as the Dutch frontier. This is a sheltered, fertile region, with rich pastures for dairy cattle on low-lying, loam-covered gravels. Towards the south are deposits of loess, giving more easily-worked soils which are mainly used for growing wheat, sugar beet and market garden crops.

The region contains both coal and lignite. There is a small coalfield round Aachen, but more important is the large lignite field west of Cologne. Most workings are in the eastern part of the field. The lignite is mined in huge open pits, which are later filled in. Almost all of West

FIG. 9.6 **The Rhine Valley** ▶

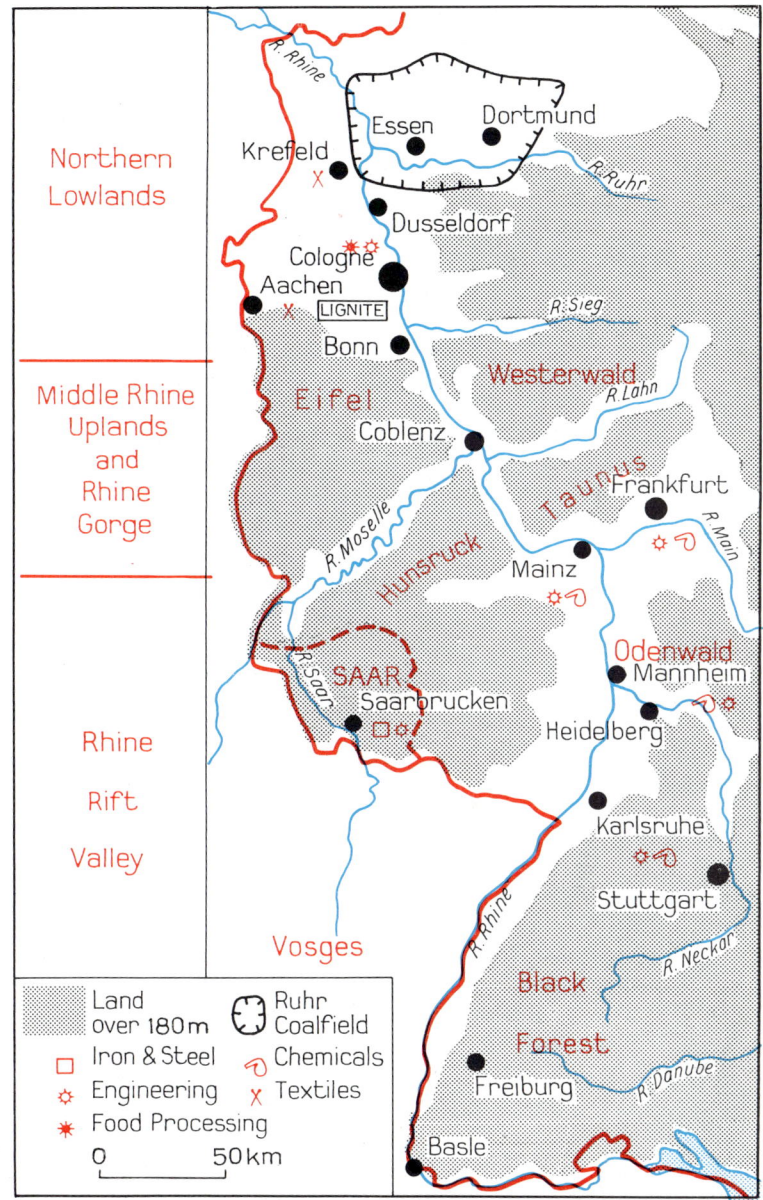

Cologne is the head of navigation for ocean-going ships, and is also an important bridge town and focus of railways. The large buildings at the bottom of the picture are international exhibition halls.

Germany's annual output of 130 million tonnes (calorific value equivalent to 40 million tonnes of bituminous coal) comes from this field. There are three main uses of the lignite – a considerable proportion is burnt in electric power stations, some is made into briquettes for domestic use, and some is used as a raw material for the chemical industry, particularly in the production of tar, lubricants and petrol.

The largest city of the Northern Lowland is *Cologne* (1,000,000), a Rhine port which can be reached by large cargo ships. It lies at a bridging point of the river and is an important railway junction. These factors, combined with its nearness to the Ruhr industrial area, have made it a major commercial and industrial centre. It has food processing, engineering and motor car industries, and is well known for its perfumes (including Eau de Cologne) and chocolate.

Textile manufacturing, with products ranging from cotton and wool to silk and synthetic fibres, is carried on in many towns including *Krefeld* and *Aachen*.

**Open-cast lignite working near Cologne** *Bundesbildstelle, Bonn*
This giant shovel-dredger digs out the lignite, which is then taken away by conveyor belt for processing.

## The Ruhr Industrial Region

The Ruhr is one of the most concentrated industrial areas in the world, with a population of some seven million living in an area measuring 80 kilometres from east to west and 50 kilometres from north to south. It has two basic industries, coal mining and iron and steel manufacture, but others have developed also, including heavy engineering, chemicals and textiles.

The Coal Measures are exposed in the Ruhr valley itself, but dip northwards, passing beneath the more recent rocks of the North German Plain. The main mining is now in the concealed coalfield area between the Ruhr and Lippe with the larger deeper mines to the north of the river Lippe. As in many British coalfields, mining is shifting to deeper seams and the new pits are much larger. The importance of the field is shown by the fact that about half the electricity used in West Germany is generated from Ruhr coal, and also that coal gas from the Ruhr is piped to towns as far away as Hanover and Heidelberg.

The presence of coking coal, together with the advantages of water transport, led to the growth of the iron and steel industry. In the early days iron ore was obtained from the Sieg and Lahn valleys, and later from Lorraine, but today most of the requirements are met by imports of high grade ore from Sweden, Brazil and other countries. Large quantities of scrap metal are also used in the steelworks.

Alongside the steel industry came the development of heavy engineering, which includes constructional work for buildings and bridges, boiler-making, and the manufacture of heavy machinery, cranes and locomotives. The main centres both for the iron and steel industry and for heavy engineering are Dortmund, Bochum, Essen, Duisburg and Dusseldorf.

The chemical industry is widely spread throughout the region, but is particularly important at Duisburg and Dusseldorf. It uses coal gas and the by-products of coke ovens to make dyes, nylon, lubricants and a host of other products. In recent years petro-chemical industries, based on the by-products of the numerous oil refineries, have developed rapidly.

The textile industry is mainly concentrated in the Wupper valley to the south, where cottons, woollens and synthetic fibres are manufactured in and around the main centre of *Wuppertal*. Also near the Wupper valley is the town of *Solingen* which has long been famous for its cutlery. In recent years coal production has declined and stands now at about 74 million tonnes (1977) a year. Also, with the introduction of the oxygen steel converter, the smaller basic and open hearth furnaces are being phased out. On the other hand a number of new industries have been introduced into the area. These include car manufacturing and branches of electrical engineering. For example, in Bochum, where pit closures caused much redundancy, car plants and varied engineering works have become established to bring back prosperity to the area.

There are four towns in the Ruhr with populations of over half a million. *Dortmund* (640,000) lies on the Dortmund-Ems canal, and is famous for its beer as well as its iron and steel. *Essen* (675,000) is often called the 'metropolis' of the Ruhr and owes much of its prosperity to the Krupp family. The Krupp combine is the largest in the region and operates coal mines, iron and steel works and engineering plants although it is no longer a family concern. *Duisburg* (590,000) lies on the Rhine and is Europe's greatest inland port. Its waterfront is now the principal steelmaking area in the Ruhr because of the advantages of low-cost imports of iron-ore. *Dusseldorf* (675,000) is another Rhine port, and serves the Ruhr as a banking and administrative centre, housing the head offices of many leading firms.

### Modern Ruhr colliery

*Bundesbildstelle, Bonn*

Coal mining in the Ruhr is highly mechanised, and output per man-shift is one of the highest in Europe. The coal, after passing through the washing and sorting sheds, is taken by conveyor belt to nearby power stations and gasworks.

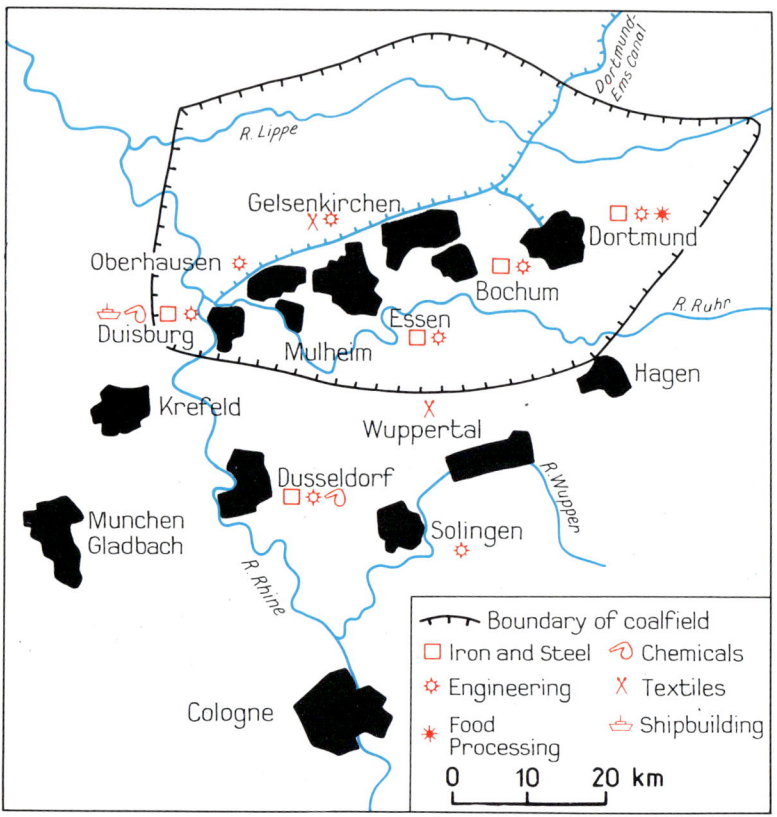

FIG. 9.7 **The Ruhr**

### The Middle Rhine Uplands and Rhine Gorge

The Middle Rhine Uplands are remnants of the Hercynian fold mountains which were worn down to a peneplain and then uplifted so slowly that the Rhine and its tributaries the Moselle, Lahn and Sieg were able to retain their courses, cutting deep gorges. The uplands were thus cut into separate plateaus, which are known as the Eifel and Westerwald in the north, and the Hunsruck and Taunus in the south. In height they generally exceed 450 metres, and are covered by thin soils with large areas of poor pasture, forest and moorland, and

94

only a little cultivated land. There is much evidence of recent volcanic activity in the Eifel in the form of volcanic cones and crater lakes, and volcanic rocks also occur widely in the Westerwald.

In contrast to the uplands, there is intensive farming in some of the narrow valleys where the rich, alluvial soils and warmer climate make possible the growing of hops, tobacco and fruit, and there are vineyards on terraced, south-facing slopes. The most spectacular scenery is along the 110 kilometre-long Rhine gorge where the river, overlooked in places by precipitous rocks crowned by mediaeval castles, presents a lively scene with its steamers and strings of barges drawn by powerful tugs.

There are a number of pleasant towns in the region, much frequented by tourists. The largest is *Bonn* (300,000), which lies on the Rhine at the northern end of the gorge, and was chosen as the capital of the Federal Republic in 1949. Another is *Coblenz*, a river port at the confluence of the Rhine and Moselle. The Moselle is canalised between Coblenz and Metz in Lorraine, and is used by barges of up to 1,500 tonnes carrying iron ore from Lorraine and coal from the Ruhr.

### The Rhine Rift Valley

Whereas the Rhine and its tributaries in the gorge section incised themselves in a rising upland, the old rocks to the south were broken by a series of north-south faults, leaving the uplands as block mountains (horsts), separated by a rift valley between 30 and 50 kilometres across.

The rift valley, between Basle and Mainz, is occupied by the Rhine, whose former meanders have now been straightened and the river confined within embankments so as to prevent flooding and improve navigation. The sandy river margins are wooded, but most of the flood plain is used as cattle pasture. Bordering the plain are river terraces which are well cultivated with wheat, maize, sugar beet and tobacco, especially where there is a covering of loess. Higher still, on the lower slopes of the uplands, are orchards, hop gardens and vineyards.

The bordering uplands, including the Vosges in the west (which lie in France) and the Black Forest and Odenwald in the east, rise steeply from the rift valley to gently rounded summits. Much of the high ground is covered with coniferous forest, but there are hill pastures in sheltered clearings, and some cultivation of hardy crops such as oats, rye and potatoes. The Black Forest reaches almost 1,500 metres above sea level, and its forests, lakes and ski slopes make it a popular tourist area.

*Frankfurt* (1,500,000) lies on the navigable river Main, is a focus of railways and motorways, and is a major growth point. It is a regional market centre and has engineering, chemical and other industries. Towns having similar functions include *Mainz, Mannheim* and *Karlsruhe. Heidelberg*, an ancient university town on the banks of the Neckar, and *Freiburg* are mainly tourist centres, with light industries.

The area known as the *Saar* became part of the Federal Republic in 1957, after a period of French control. It is wooded hill country with deeply incised rivers, and contains a small coalfield. Its importance has for long been the iron and steel industry, the ore being

**Crater lake in the Eifel**   *Deutschen Zentrale fur Fremdenverkehr, Frankfurt*

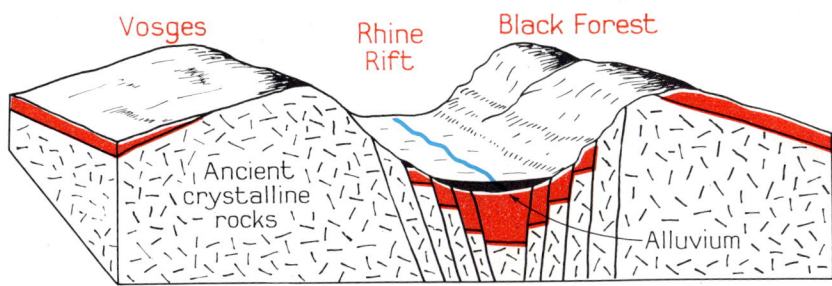

FIG. 9.8 **Simplified block diagram of the Rhine Rift Valley**

obtained from nearby Lorraine. Recently the manufacturing of cars, electrical goods, fertilisers and other chemical products has developed, particularly in the vicinity of Saarbrucken, to give new life to the region.

## The Central Uplands

The Central Uplands occupy much of central and southern Germany to the east of the Rhineland, and are a region of great physical diversity resulting from a wide variety of rocks and geological structures. The uplands rise to between 300 and 900 metres above sea level, and are well-forested with beech and oak trees on the lower land, and conifers including spruce at higher levels. The forests provide employment in wood cutting, sawmills and woodworking factories. The soils are generally infertile, but there are considerable areas of cleared land which are used for cattle pasture.

The valleys and basins, on the other hand, contain fertile soils and there is livestock farming as well as the growing of wheat, oats, rye and potatoes. Rural over-population in the past has led to the development of cottage industries including woodcarving and the making of toys, furniture, musical instruments, clocks and watches. Many of these are still carried on but most of the work is now done in factories. The larger towns are regional market centres serving the lowlands in which they lie, and many have large-scale specialised industries.

The lack of resources is partly made up for by the flourishing tourist industry, as the hills and forests make excellent walking country, and many of the smaller towns have preserved their mediaeval character and attract large numbers of visitors.

In the north is the upper Weser basin, which is almost surrounded by hills, including the isolated Hercynian horst of the Harz Mountains in the east, the volcanic hill masses of the Vogelsberg and Rhon in the south, and the Middle

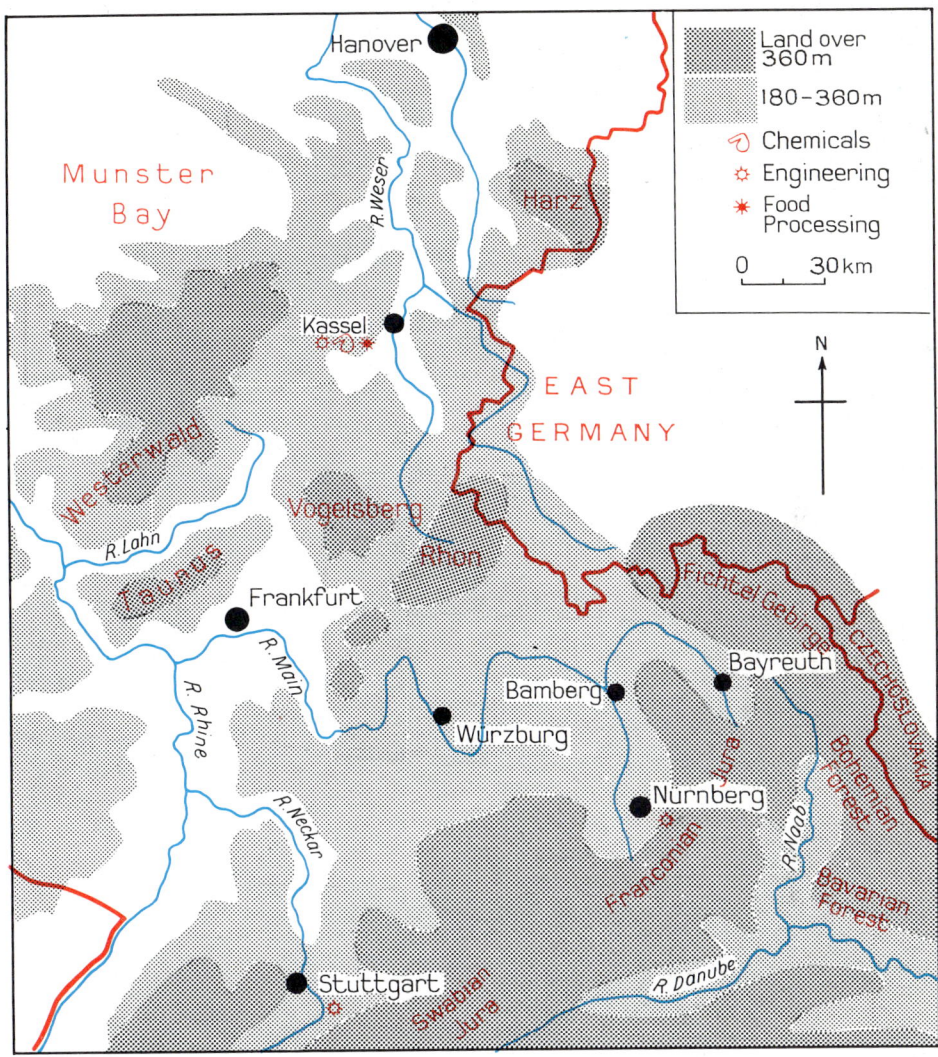

FIG. 9.9 **The Central Uplands**

**Heidelberg**                                      *J. Allan Cash*

This historic university town lies at a bridging point of the Neckar, among wooded hills.

Rhine Uplands in the west. There is little natural wealth apart from timber and some scattered mineral deposits. The working of lead, zinc and copper was formerly important in the Harz Mountains and is still carried on to a small extent. The Weser basin is the most important farming area, and *Kassel*, which lies at the head of navigation of the river, is the only large town. It has a number of industries

**Rothenburg**                                      *J. Allan Cash*

One of the mediaeval walled towns of Bavaria, which are now great tourist attractions.

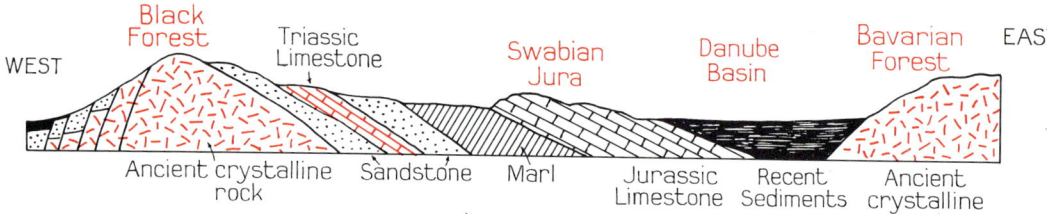

FIG. 9.10 **Simplified section across the southern part of the Central Uplands**

**The Black Forest**                   *Bundesbildstelle, Bonn* ▶

(exercise on page 101)

including railway engineering, food processing and chemical manufacture. It is a major route centre between north and south.

Farther south-east are the uplands of the Fichtel Gebirge, Bavarian Forest and Bohemian Forest, the latter lying along the Czech frontier. Furniture-making and glass-making are important in the Bohemian Forest. For glass-making, use is made of the local pure quartz sand, wood ash to provide the necessary potash, and wood as fuel. The nearby Naab valley contains some workings of brown coal and iron ore which have influenced the development of small metallurgical industries.

Much of the country between the Rhineland and Bohemian Forest (north of the Danube) is formed of younger rocks including limestone, sandstone and clay. These rocks generally dip to the south-east, so that the harder layers stand out as north-west facing scarps. The most prominent of these are the Swabian and Franconian Jura which cross the region from south-west to north-east. They consist of permeable limestone which accounts for the bare, dry landscape, but on the lower dip slopes clay appears at the surface, providing some useful farmland.

The most important farming areas are the Main and Neckar basins whose warm climate and fertile, loess soils favour the growing of wheat, sugar beet, hops and tobacco. Fruit orchards and vineyards do well on the valley sides, and there is cattle pasture on the damper soils of the valley floors. The Main and Neckar have been made navigable by the construction of weirs and locks, and the good communications along the valleys have encouraged industrial development. *Stuttgart* (1,900,000 in the city region) on the river Neckar, is a route focus and a financial and commercial centre. Among its industries is the large Mercedes-Benz car works. *Nurnberg* (520,000) is another large town and is concerned with the manufacture of machinery, diesel engines and electrical goods.

## The Bavarian Foreland and Alps

In the south, along the Austrian frontier, is the rugged limestone range of the Bavarian Alps, forming only the narrow northern fringe of the main Austrian Alps. Here is Germany's highest mountain, the Zugspitze (2,963 metres) and other impressive peaks, and a number of attractive lakes. This is one of the country's most popular tourist regions and contains considerable reserves of timber.

North of the Bavarian Alps is the Bavarian Foreland, a broad plateau sloping gently down to the Danube. Much of the plateau is covered by glacial deposits and its surface has been dissected by streams which have formed wide valleys. In the south the surface consists of stony clay, gravel and peat, giving poor soils. Much of the land is under permanent pasture, grazed by cattle, with stretches of heath and forest on the morainic ridges. In the north some of the land is loess-covered, and there is more arable farming, including the growing of oats, rye, barley, potatoes and fodder crops. The soils are especially fertile along the Danube valley, where fruit, hops and market garden crops are important.

Generally the Foreland is a thinly peopled area, dominated by the city of *Munich* (1,500,000 with suburbs). Munich lies where an important route from north Germany to Italy crosses an east–west route from Paris to Vienna. It was the old capital of Bavaria, but its present importance is derived from its commerce and industries, the latter including the manufacture of textiles and chemicals, food processing, brewing, engineering and motor-cars. *Augsburg* is another large town with textile and engineering industries, using hydro-electric power from the Lech.

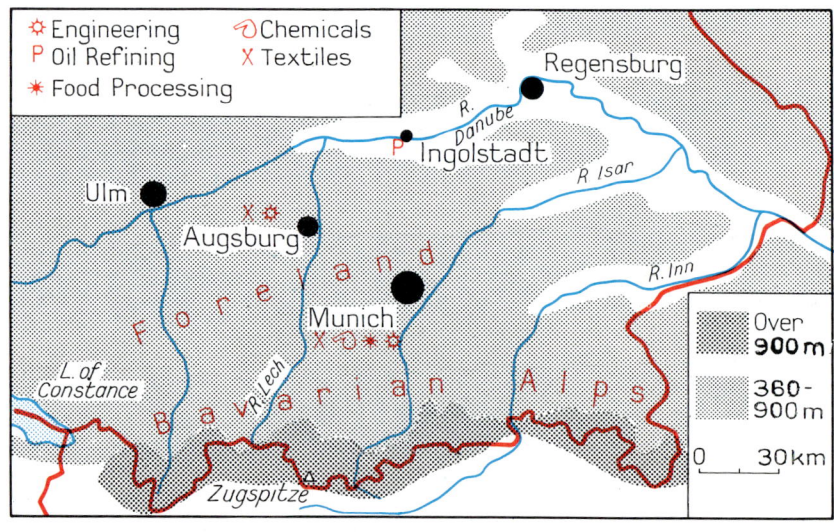

FIG. 9.11 **The Bavarian Foreland and Alps**

**Bavarian Alps, with the Zugspitze**

The varied physical features of West Germany are illustrated by the photographs on pages 88, 96, 99 and on this page. Describe each scene and comment on the geological processes which are responsible for the land forms.

## Exercises

*Answer in note form :*

1. Why is West Germany not self-supporting in foodstuffs?
2. What geographical factors have favoured the development of industry in the Ruhr?
3. Locate an example of each of the following in West Germany, and explain their formation: horst, gorge, rift valley, volcanic area.
4. Which are the principal tourist areas of West Germany, and what have they to offer to the visitor?

*Essay Questions*

1. With reference to the ports of Bremen and Hamburg, (a) draw one sketch map to locate both ports, (b) what geographical factors have led to the growth of each port?, and (c) Show how and why one is more important than the other.
2. Describe the main effects of glaciation on the surface of the North German Plain and state how these have influenced the type of farming.
4. How can a knowledge of the geography of West Germany help to explain the great concentrations of population shown in Figure 9.12?

### Southern Bavaria

*Bundesbildstelle, Bonn*

This road connects Munich with Innsbruck in Austria, and the view is to the south, across the River Isar and towards the Bavarian Alps.

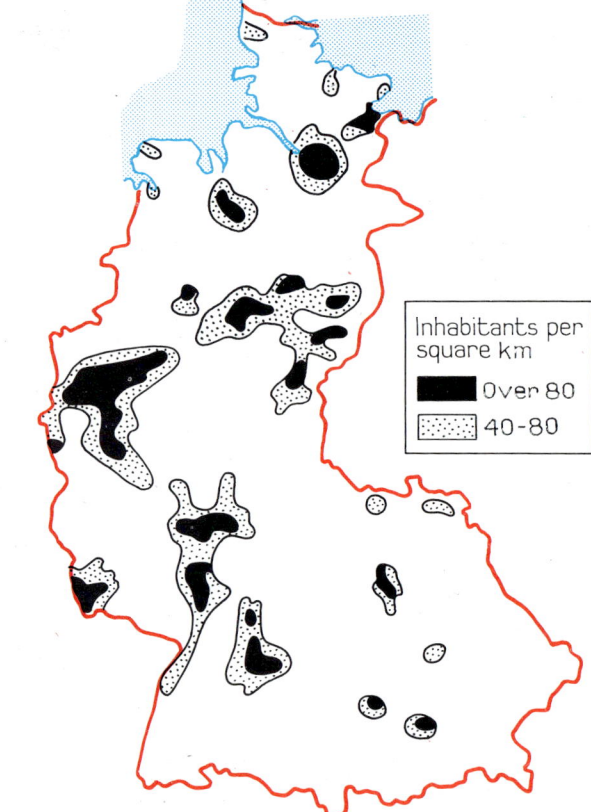

FIG. 9.12 **Major concentrations of population in West Germany**

# 10: THE GERMAN DEMOCRATIC REPUBLIC (East Germany) AND BERLIN

East Germany has an area of 106 square kilometres and a population of 17½ million, and is a communist state, closely allied with the U.S.S.R. and other east European countries.

After the war there was a large influx of refugees from the former German lands and from Czechoslovakia, but since then there has been an even greater loss of population to West Germany. The new state faced considerable difficulties, for it lacked the natural resources of West Germany and its industries were greatly handicapped by the loss of managers, engineers and skilled technicians who had fled to the west. Nevertheless great progress has been made in re-building the economy, and the standard of living of the people, which was at one time far below that of West Germany, has shown a marked improvement in recent years.

By far the greater part of East Germany lies on the North German Plain, whose undulating surface is largely covered by glacial material, including widespread deposits of boulder clay, sand and gravel, with wind-borne loess along the southern borders. In the south-western and southern parts of the country the land rises to the wooded Hercynian uplands of the Harz Mountains, Thuringian Forest and Erz Gebirge.

The climatic conditions are slightly more continental than those of West Germany. Winters are more severe and last longer, and in summer there are more hot spells, which often end in thunderstorms. Precipitation is lower and shows a more marked summer maximum. As an example, Berlin has a mean January temperature of −2°C, a July mean of 18°C, and an annual precipitation of 585 mm.

## Agriculture

Farming is generally less productive than in West Germany, mainly because of the larger proportion of infertile land on the glacial sands and gravels. On this poor land rye and potatoes are grown, and pigs and cattle are numerous. The loess areas in the south provide more fertile soils on which wheat and sugar beet are grown. In 1960 collective farming was introduced in place of the peasant small-holdings. Each collective farm covers tens of thousands of hectares and is worked along co-operative lines, with the farm income being shared out as wages in accordance with the amount of work done. The new system did not work well at first, probably because of the loss of personal incentives, and output declined. However it is now steadily improving as the greater efficiency of mechanised, scientific methods starts to bear fruit.

## Industry

Industrial development has been given great priority, but East Germany is lacking in many raw materials. There is very little good quality coal or iron ore, but on the other hand there are large lignite deposits, as well as deposits of potash, salt and uranium. Before the war eastern Germany was well known for its high quality textiles, china, optical equipment, precision tools and musical instruments, all products of skilled workmanship, but there was very little heavy industry.

One of the first objectives of the East German government was the creation of a large iron and steel industry, making use of iron ore

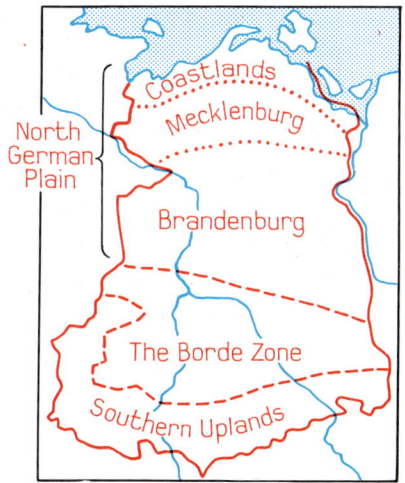

FIG. 10.1 **The regions of East Germany**

from the Ukraine and coal from Polish Silesia. Modern steelworks have been constructed at the new town of Eisenhüttenstadt and at a number of other centres, and production has now reached nearly 7 million tonnes per annum. Power supplies for industry are derived mainly from local lignite, and also from mineral oil and natural gas brought by pipeline from the Soviet Union. In addition to the iron and steel industry there has been a considerable development of the heavy engineering, electrical and chemical industries, and particular attention is now being paid to expanding traditional East German manufactures, for example photographic and optical equipment.

East Germany is a member of the *Comecon* (the Council for Economic Co-operation), which is the communist equivalent of the Common Market, and three quarters of its trade is with other Comecon countries. The U.S.S.R. supplies most of the raw material requirements and considerable quantities of foodstuffs, and is the principal market for its exports, 85 per cent of which consist of manufactured goods.

## THE REGIONS OF EAST GERMANY

### The North German Plain

The Baltic coast of East Germany is flat, with numerous indentations. There were only small ports here before 1939, since most goods from the industrial regions to the south were shipped from Hamburg, Bremen and Stettin (now the Polish port of Szczecin). Since 1945 *Rostock* has been built up into a major port with a motorway linking it to Berlin. It has large shipyards where medium-sized passenger and cargo ships are built by highly automated methods, and is also East Germany's principal fishing port. Considerable harbour improvements have also taken place at *Wismar*.

Inland is a low-lying plain of boulder clay, a fertile area with most of the land under crops including sugar beet, rye, potatoes and fodder crops for feeding to cattle and pigs. The land rises southwards to the lake-studded plateau of Mecklenburg, which is built of glacial sands and gravels and is crossed by terminal moraines of very coarse material. This is an area of poor soils and is thinly peopled, with considerable stretches of pine and beech forest.

Farther south, in Brandenburg, much of the ground is covered by sandy outwash material on which there are extensive pine forests. This area is crossed by a number of broad, swampy valleys running

FIG. 10.2 **East Germany** ▶

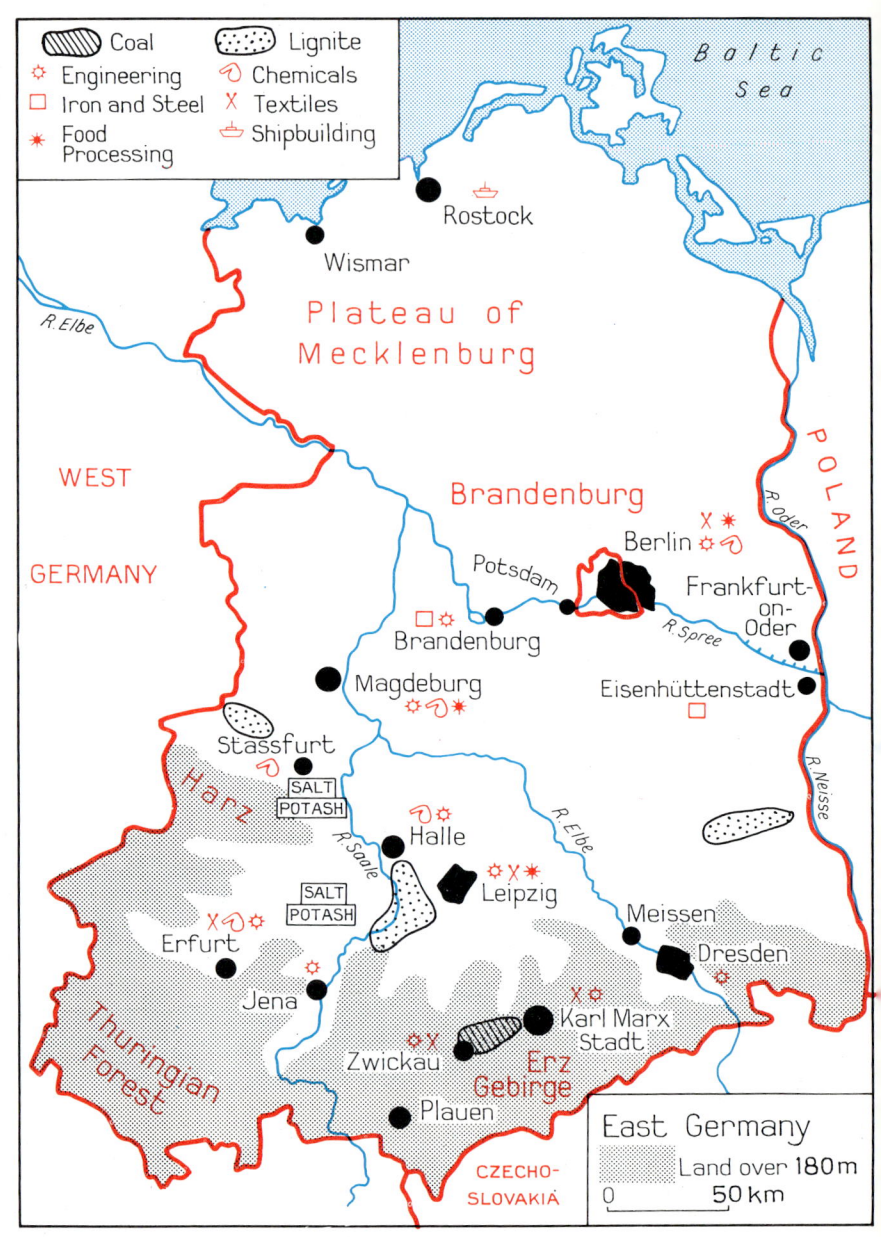

from east to west which were carved by melt-water flowing westwards from the edge of the ice sheets. These swampy valleys are known as urstromtaler and their floors have been drained to give fertile alluvial soils which produce good crops of wheat, sugar beet and tobacco. In the vicinity of Berlin (which lies on an urstromtaler) there is intensive fruit and vegetable growing, both in the open and under glass.

*Brandenburg* is one of the centres of the newly-developed steel industry, using scrap as the raw material, and also has a large tractor works. *Frankfurt-on-Oder* is a river port on the Polish frontier, and south of it, on the Oder-Spree Canal, is the new town of *Eisenhutten-stadt*, which has the most modern iron and steel works in East Germany.

## Berlin

Berlin lies at a crossing point of the River Spree and is the largest city in all Germany. Although it possessed no great natural advantages, it became the capital of Brandenburg, then of Prussia, and finally in the nineteenth century of a unified Germany. Its position near the centre of the North German Plain made it a great focus of rail, road and water routes, and this in turn led to the growth of industries.

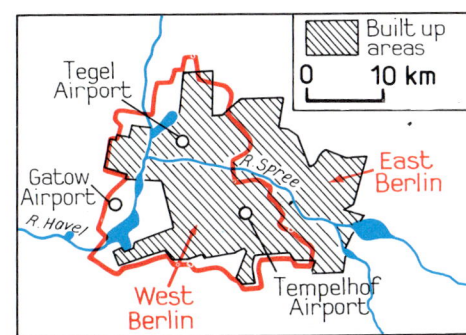

FIG. 10.3 **Berlin**

**West Berlin** (above right)    *Presse und Informationsamd der Bundesregierung* ►

**East Berlin** (right)    *ADN/Zentralbild* ►

The centre of Berlin was three-quarters destroyed during the war, and a systematic rebuilding programme is under way in both eastern and western sectors.

The city was greatly damaged at the end of the Second World War, and in 1945 it was divided into four sectors by the occupying powers. Today *West Berlin*, the portion occupied by the three western powers in 1945, remains an uneasy island surrounded by East German territory and yet bound, politically and economically, to West Germany. It has a population of $2\frac{1}{4}$ million, and in spite of handicaps to trade due to its isolation, it has developed into a flourishing city with numerous industries including the manufacture of electrical equipment, precision instruments, clothing and foodstuffs. *East Berlin* has a population of $1\frac{1}{4}$ million and is the capital of East Germany. It also has important industries, notably the manufacture of electrical equipment and chemical products, as well as engineering. A short distance to the south-west of Berlin is the historic town of Potsdam, with its royal palaces.

## The Borde Zone

To the south of the North German Plain is a broad strip of lowland which follows the northern edge of the Hercynian uplands. Here the solid rock is covered with a mantle of loess, giving soils of considerable fertility. It is the best farm land in East Germany, particularly for wheat, sugar beet, potatoes, pigs and cattle, and in some sheltered valleys on the edges of the hills, orchards and even vineyards flourish.

The zone is broadest in the Leipzig Bay and extends north as far as Magdeburg. The Leipzig Bay is the most important industrial area in the country, and contains vast deposits of lignite which, together with a second major field in Lusatia, west of the Neisse, make East Germany the largest producer of this fuel in the world. About 250 million tonnes are mined each year from open cast workings. The lignite has a much lower calorific value than hard coal, and tends to disintegrate when transported for any distance. Consequently much is used locally in electricity power stations, and some is made into briquettes and metallurgical coke.

There are also large deposits of potash and salt near Stassfurt and Halle, and these are used, together with lignite, as raw materials for the chemical industry, which produces fertilisers, synthetic fibres, soap, paper and glass.

The region is dominated by *Leipzig* (575,000), which suffered widespread destruction as a result of war-time bombing but has gradually regained its importance. Among its many industries are engineering and the manufacture of textiles and musical instru-

**Halle**

*ADN/Zentralbild*

Halle is an industrial town, but has retained its ancient market square, with the 16th century Red Tower on the right. It was the birthplace of Handel, and an annual Handel Festival is held here.

ments. Before the war it was Germany's main centre for printing and publishing, and these are still carried on despite the fact that many of the leading firms moved to West Germany after the war. Leipzig is well-known for its historic twice-yearly industrial fairs, where manufacturers from both eastern and western Europe exhibit their goods. *Magdeburg* (275,000) is the main commercial and industrial centre in the northern part of the Börde zone, and its manufactures include food products, heavy machinery and chemicals. Other towns in the region are *Halle* (250,000), which has chemical and engineering industries, *Jena*, where the famous Zeiss optical instruments are produced, and *Erfurt*, with chemical, engineering and clothing industries.

◄ **Zeiss Works at Jena**    *VEB Carl Zeiss Jena*

The old university town of Jena lies in the valley of the River Saale and is overlooked to the east and west by wooded mountains. The works were founded by Carl Zeiss in 1846, and are the largest in the world for precision mechanics and optics, employing 22,000 people.

**Eisenhuttenstadt**    *Deutsche Bauinformation*

This is a completely new town and lies on the west bank of the River Oder, where the Oder-Spree Canal leaves the river. Its giant steelworks, using Polish coal and iron ore from the Ukraine, have been an important factor in East Germany's industrial development.

## The Southern Uplands

The Harz Mountains and Thuringian Forest are fault-bounded plateaus (horsts) rising to over 1,000 m above sea-level. They are well-forested but contain clearings which provide pasture for livestock. Formerly there was a large mining population in the uplands, and as the mines were worked out other industries developed, mainly specialised industries which required a great deal of skill and little raw material, for example the making of toys, glassware, musical instruments and fine metal articles. Since the war there has been a revival of copper mining in the eastern foothills of the Harz.

The Erz Gebirge (Ore Mountains) form the boundary between East Germany and Czechoslovakia. They were also the scene of great mining activity which has now ceased apart from the mining of uranium, the chief source of nuclear energy, which has been greatly increased since 1945. The northern slopes form a dissected plateau known as the Saxon Uplands, and here is a small coalfield providing the only hard coal in East Germany. For many centuries this has been an important manufacturing region particularly for textiles. Originally there was a flourishing domestic woollen industry, the wool coming from local sheep. Later came the manufacture of cotton, linen, jute and more recently, artificial fibres. The industry is widely dispersed in many small towns, with *Karlmarxstadt* (formerly Chemnitz) and *Zwickau* as the main centres. These towns also have engineering industries including the manufacture of motor vehicles.

The main commercial centre of this industrial region is *Dresden* (510,000) which lies on the Elbe just below where the river breaks through the mountain rim forming the frontier with Czechoslovakia. Its principal industries are the manufacture of iron and steel, precision instruments, machine tools and electrical goods. Not far away is the town of *Meissen*, where the famous 'Dresden' china is made.

## Exercises

*Answer in note form:*

1. Describe the East German boundary from Rostock, in a clockwise direction, mentioning the physical features which form the frontiers.
2. Why has Rostock been developed into a major port?
3. Compare East Germany's climate with that of West Germany.
4. Locate East Germany's main centres of steel production.
5. Compare the systems of farming in East and West Germany.

*Essay Question*

Describe, with the aid of a sketch map, the manufacturing industries of East Germany, and discuss the problems of raw material and fuel supplies.

**Saxon Switzerland**  *Lex Hornsby and Partners*

This area is situated in the south-east corner of East Germany and is very popular with walkers and climbers. The strongly-jointed sandstone rocks have been weathered into unusual shapes by the action of wind, rain and frost.

# 11 : SWITZERLAND

Switzerland has an area of only 41,000 square km, and is a mountainous country, wedged between France, Germany, Austria and Italy. Of its population of approximately 6½ million, 74 per cent speak German, 20 per cent French, 4 per cent Italian and 1·3 per cent Romansch, a language related to Latin. German is spoken in the north, east and centre of the country, French in the west, and Italian in the south. Switzerland is a federal state composed of twenty-two cantons, each with a considerable degree of autonomy, and in its foreign policy it has followed a traditional role of neutrality between the great powers.

The country is bounded on all sides by natural frontiers: to the east and north by the Rhine, to the north-west by the Jura Mountains, and to the south by the mighty peaks of the Alps and by Lake Geneva. Only the canton of Ticino and a few valleys of eastern Switzerland lie on the southern slopes of the Alps. Between the Rhine and Jura in the north, and the Alps in the south, lies the Swiss Plateau, a region which has played a great part in the development of the Swiss economy.

The climate is of temperate interior type, but the country's position fairly close to the Atlantic seaboard of Europe also makes it subject to maritime influences. As a result, the summers are warm and showery, whilst the winters are not too severe on the Plateau where snow seldom lies for long. Bern, which lies at about 600 m above sea level, has a January mean temperature of —1°C, a July mean of 20°C, and an annual rainfall of 810 mm. Conditions in the mountains are more severe and precipitation is much greater, including heavy snowfall in winter.

Switzerland is not rich in natural resources. It is without coal, mineral oil and natural gas, has few metal ores, and only 27 per cent of the land is suitable for farming. Transport costs are high because of the lack of a sea coast, and because roads and railways are expensive to build and maintain in such a mountainous country. In spite of these disadvantages the Swiss, by making the most of their limited resources, have attained a high level of prosperity.

### Agriculture

Agriculture occupies an important place in the economy of the country. Swiss farms are very small, averaging some 7 hectares; in many places the relief of the land prevents the use of large machines. Nevertheless methods are scientific and intensive, and farmers are able to supply the nation with three-fifths of its food requirements. A considerable variety of temperate crops and fruits are grown, but cattle are the chief source of income for the farmer, followed by pigs. The Swiss Plateau is the most important farming region, and there is some farming also in the lower valleys of the Alps and Jura.

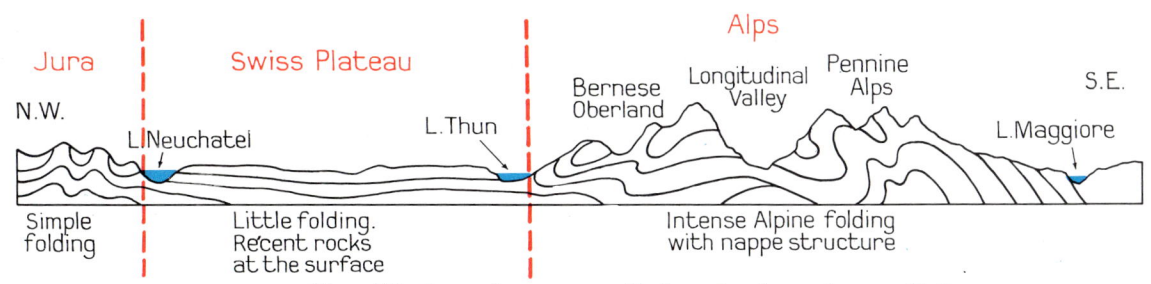

FIG. 11.1 **Simplified section across Switzerland to show relief and structure**

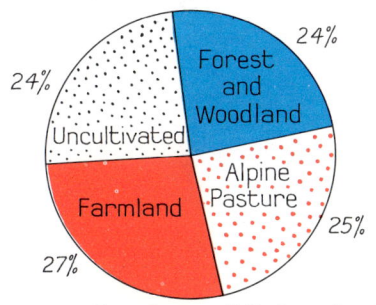

FIG. 11.2 **Land use of Switzerland**

## Power and Industry

Power supplies are derived from local hydro-electric resources, and also from imported mineral oil and coal. The heavy rainfall of the mountain areas and the numerous lakes and waterfalls provide excellent conditions for the construction of hydro-electric stations. The electricity is consumed in factories, by the railways and in the home. The available resources of water power will soon be fully utilised, and already a beginning has been made in using other sources of power for generating electricity, including an oil-fired power station and a nuclear power station. There are other projects for nuclear power stations to be built during the coming years.

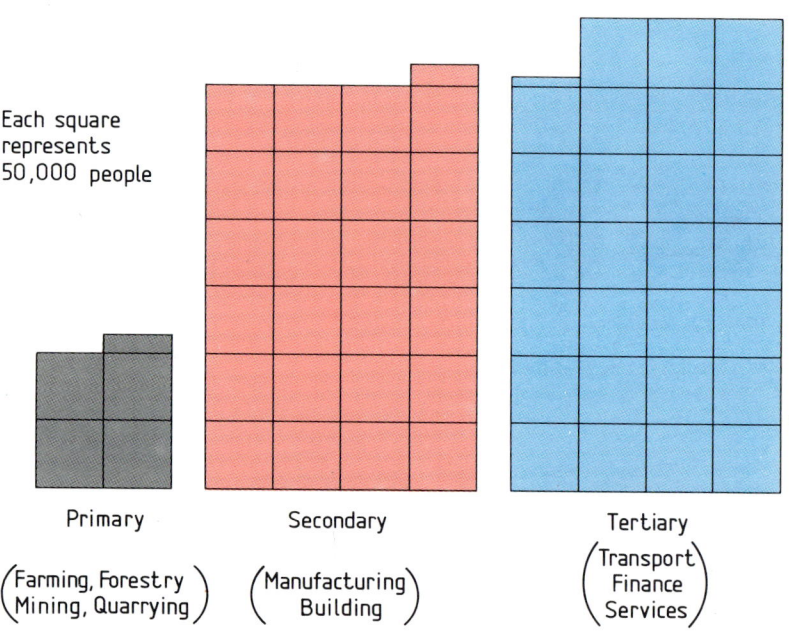

FIG. 11.3 **Main sectors of employment in Switzerland**

In industry, because of the high cost of transport of raw materials, there is a concentration on products which are valuable in proportion to their bulk, and which require a high degree of skill and specialisation. Engineering employs the largest number of industrial workers and its varied products include electrical equipment, machine tools and precision instruments. Switzerland is also the world's largest producer of clocks and watches. The manufacture of textiles and clothing, chemicals and food processing are carried on in many parts of the country.

The magnificent Alpine scenery and the winter sports facilities attract large numbers of foreign tourists. The Swiss excel at hotel-keeping and have built many resorts, both on the lakes and in the mountains, as well as fine roads, mountain railways and cable cars to bring the most spectacular scenery within easy reach of the visitor.

The Swiss economy relies heavily on foreign trade. Among the imports are mineral oil, coal and other industrial raw materials, foodstuffs and animal fodder, and a wide range of manufactured goods including machinery and motor vehicles. The principal exports are machinery, precision instruments, clocks and watches, chemical and pharmaceutical products, textiles and shoes. The value of the exports is not sufficient to meet the cost of the imports, and the balance is covered by receipts from tourists and charges for transport, banking and insurance services.

## THE REGIONS OF SWITZERLAND

### The Alps

Over half of Switzerland lies in the Alpine region, which consists of two main ranges, each running approximately from south-west to north-east. The southern range, much of which forms the frontier with Italy, is of crystalline rocks. It includes the Pennine Alps (with Monte Rosa, 4,638 m and the pyramid-shaped Matterhorn, 4,505 m, and the St Gotthard Massif, and becomes lower in the east where it is broken by valleys, notably the Engadine, drained by the Inn. The northern range is mainly of limestone and is known in the west as the Bernese Oberland (Finsteraarhorn, 4,274 m and Jungfrau, 4,166 m). Between the northern and southern ranges is a narrow longitudinal valley occupied by the upper courses of the Rhine and Rhône. There are a number of passes into Italy, including the St Gotthard and Simplon which are used by trans-European railways, and to the south of the St Gotthard lies the Italian-speaking canton of Ticino with parts of Lakes Maggiore and Lugano.

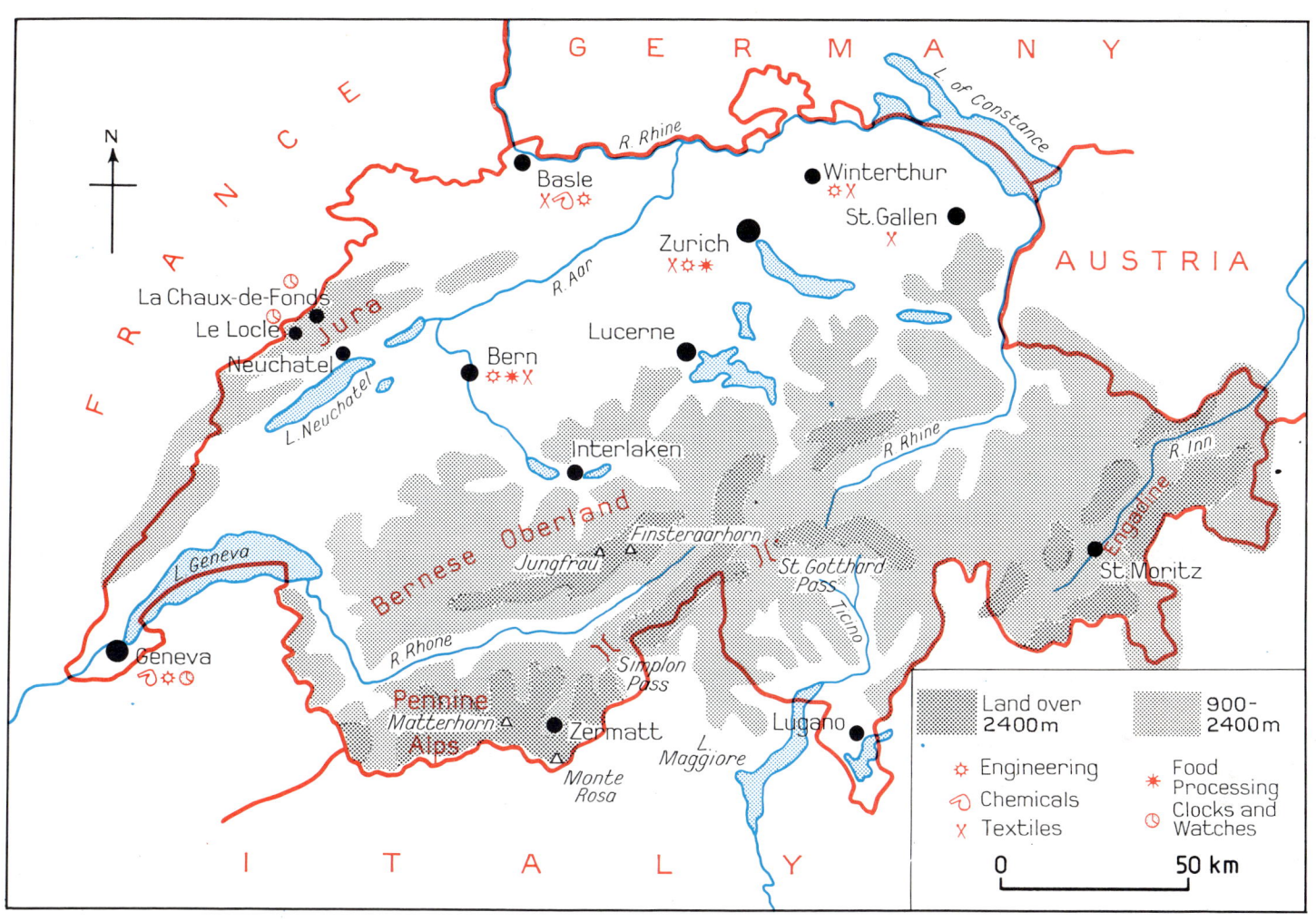

FIG. 11.4 **Switzerland**

**Alpine glacier**

Make a line drawing or plan of the glacier, labelling one example of each of the following: pyramidal peak, cirque, arête, lateral moraine, medial moraine.

Compare the work of the glacier and the river (in the previous photograph) in the shaping of their valleys.

*Swiss National Tourist Office*

113

**Hydro-electric power station**

*Swiss National Tourist Office*

Why are the Swiss Alps suitable for the development of hydro-electric power? Explain briefly how the power is generated, and state the advantages of hydro-electricity over other sources of power.

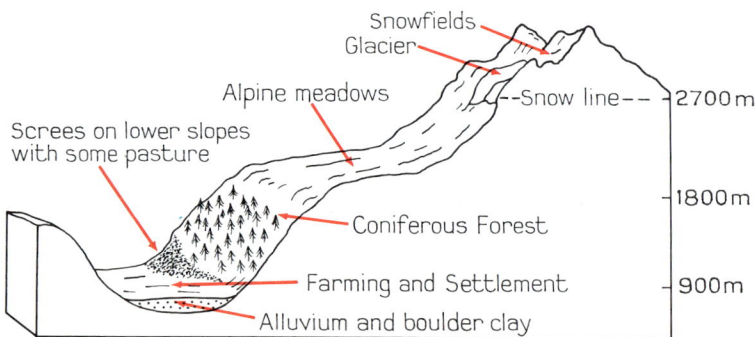

FIG. 11.5 **Block diagram of a portion of an alpine valley**

The mountain slopes are covered in coniferous forests up to about 1,800 m, and above the forests are Alpine meadows lying on broad shoulders of land and reaching up to the snow line at about 2,700 m. The Alpine summits rise out of vast snow fields, from which glaciers descend towards the main valleys. The influence of past glaciation is apparent in the pyramidal peaks, arêtes, cirques and deep, U-shaped valleys, some of which contain lakes.

The small population of the Alpine region is concentrated in the valleys, where hay, roots and hardy cereals are grown. Cattle are the mainstay of the farmer and transhumance is practised. The cattle are driven to the Alpine meadows in early summer. The herdsmen stay with them, living in wooden houses where the milk is made into butter and cheese. In autumn the cattle return to the valley farms where they spend the winter in barns, and are fed on hay and other stored fodder. Employment is also provided by the forests, the timber being used for fuel, for building chalets, and for wood carving.

Some valleys are much affected by a warm wind called the Föhn,

which blows sometimes in winter over the Alps from south to north, towards a depression on the north side. As the air descends it is warmed by compression and the temperature rises rapidly. Fruit-growing is often made possible by the warming effect of the Föhn, but the sudden melting of the snow which it produces may cause avalanches.

There are many hydro-electric stations in the Alps. Water is stored in high level lakes and reservoirs, and the power stations are sited in the valleys. Power is conveyed by high tension cable to all parts of Switzerland, but some is used for aluminium smelting and for the manufacture of heavy chemicals and fertilisers in the upper Rhône valley.

The Alps are also the country's main tourist region, and there are numerous resorts which offer accommodation to the visitor, both in summer and winter. Some of the largest resorts, such as *Lucerne, Interlaken* and *Lugano*, lie on lakes to the north and south of the main Alpine ranges, but there are others including *Zermatt* and *St Moritz* which are situated in the mountains.

## The Swiss Plateau

The Plateau, which covers about one third of Switzerland, varies in height from about 900 metres near the Alps to 400 metres near the Rhine valley and Jura. It consists of recently-formed rocks, concealed by deposits of boulder clay and coarse morainic material left by an ice sheet which covered the area during the Ice Age. There are many hills of morainic origin, and the blocking of the valleys by glacial drift led to the formation of numerous lakes, which drain to the Aar or Rhine.

About two-thirds of Switzerland's population live on the Plateau. Agriculturally it is the most important part of the country. The warm, rainy summers are ideal for the growth of meadow grasses and fodder crops, and dairy cattle, beef cattle and pigs are kept in large numbers. Some land is also under wheat, sugar beet, tobacco and vegetables, and there are vineyards especially on the south-facing hill slopes north of Lake Geneva and north-west of Lake Neuchatel. Fruit trees, mainly apples, pears and black cherries, are numerous too.

The Plateau is also by far the most important manufacturing region. The industries are widely dispersed since they make use of hydro-electric power which can be transmitted easily to wherever it is needed. The leading industry is engineering, whose products range

**Lucerne**                                   *Swiss National Tourist Office*
This is Switzerland's largest lake resort. The view is of the old town along the Reuss river, with the ancient water tower.

◄ **Typical scene on the Swiss plateau**   *Swiss National Tourist Office*

The lower land is generally flat and fertile, being largely covered by boulder clay and silt, and the warm summers encourage intensive farming. Many of the hills are of morainic origin and have infertile soils, so that they have been left under woodland.

from hydro-electric turbines and heavy machinery to sewing machines and typewriters. The main centres of the engineering industry are *Bern, Zurich, Geneva* and *Winterthur*. The chemical industry is represented by the manufacture of dyes and pharmaceuticals at *Basle*. Textile and clothing manufacture is especially important at *St Gallen, Zurich* and *Basle*. The processing of farm produce is carried on throughout the region, and includes the manufacture of butter, cheese, condensed milk, chocolate, jams, soups and meat extracts.

The country's major cities belong to this region. *Zurich* (745,000) is Switzerland's main financial and commercial centre and has varied industries including engineering, textiles and leather. *Bern* (280,000) is the federal capital and occupies a defensive site within an incised meander of the Aar. It is a route centre and has engineering, clothing, printing and chocolate industries. *Geneva* (320,000) lies on Lake Geneva, where the Rhône leaves the lake. It is the chief town of the French-speaking part of Switzerland, and houses the headquarters of a number of world organisations including the International Red Cross and the World Health Organisation. It has engineering, watch and jewellery industries. *Basle* (530,000) is a Rhine port and a meeting point of routes from Germany (along the Rhine), France (via the Belfort Gap) and Italy (via the Alpine passes). Its port handles one quarter of Switzerland's foreign trade, and it has engineering, textile and chemical industries.

## The Jura

The Swiss Jura are a classic example of simple folding. The limestone of which they consist forms a series of parallel ridges and valleys, trending from south-west to north-east. Most of the ridges, which rise to over 1,500 m, are formed by the anticlines, and the valleys by the synclines. Rivers follow longitudinal courses in the valleys, but some turn sharply to cut through the ridges in deep gorges, called cluses. Others pass into underground cave systems.

The slopes are forested, with deciduous trees (especially oak and beech) on the lower land and conifers above. Forestry and woodwork-

ing industries have developed in many areas. Dairy farming is important in the valleys, and much of the milk is used in the manufacture of chocolate, Gruyère cheese, and condensed milk.

The main centres of the Swiss clock and watch industry are in the Jura region. It first began as a domestic industry, offering employment in the winter when there was little work on the land. Today it is a highly-organised factory industry, carried on in a number of towns including *Le Locle* and *La Chaux-de-Fonds*.

## Exercises

*Answer in note form:*
1. Compare the structure of the Alps and Jura.
2. What kinds of farming are possible in the Swiss Alpine region?
3. In what ways has Swiss industry adapted itself to the lack of local raw materials?
4. Why do two-thirds of Switzerland's population live on the Swiss Plateau?

*Essay Question*
Describe the effects of glaciation on the physical and human geography of Switzerland.

**Basle**   *Swiss National Tourist Office*

The harbour is always a scene of great activity, since a large proportion of Switzerland's imports arrive here by barge. This is looking down-river, with France on the left bank and Germany, farther downstream, on the right bank.

# 12: AUSTRIA

For many centuries Austria was the nucleus of the great Austro-Hungarian Empire, and its capital, Vienna, was one of the most glamorous cities in Europe, but following the collapse of that empire at the end of the First World War, it was reduced to a small republic with only limited resources.

Today it has an area of 83,000 square km, and a population of 7·5 million. Over two-thirds of the country is mountainous, with a series of high Alpine fold mountain ranges in the west and south, similar in character to those of Switzerland. To the north and north-east the mountains and valleys fall away to the River Danube and the Vienna Basin, the only large lowland in the country.

Austria's climate is of temperate interior type, and is only slightly affected by maritime influences. Conditions on the lower land are typified by the figures for Vienna, which has fairly warm summers (mean July temperature 22°C), cold winters (mean January temperature −2°C), and an annual rainfall of about 600 mm. In the mountains the climate is more severe, with much greater precipitation including heavy snowfall in winter.

## Agriculture and Lumbering

Farming employs about 12 per cent of the working population, mainly on small family farms. Barely half of the country can be used for food production, but every effort is made to extract the utmost from the land and the farmers are able to provide nearly 90 per cent of the nation's requirements. The most important agricultural areas are on the alluvial soils of the Danube valley and on the lower land in the east, including the Vienna Basin, where the warm summers permit the cultivation of wheat, maize, sugar beet, tobacco and fodder crops. Cattle and pigs are widely kept, and vineyards and fruit orchards are found in some areas. In the mountain regions cultivation is mainly restricted to the valleys where hay and oats are grown for feeding to livestock. The Alpine pastures are used for summer grazing.

Forestry also provides considerable employment in the mountains, where there are large reserves of coniferous timber. The timber is used for house-building, furniture, wood-carving, fuel, and in pulp and paper mills, and large quantities of timber and wood pulp are exported.

## Power and Industry

Austria has adequate power resources. Its mountain lakes and valleys lend themselves admirably to hydro-electric installations, and since the war many new schemes have been undertaken, both in the Alps and on the Danube and its tributaries. As a result of these, Austria now has a surplus of electricity and some is exported to nearby parts of Germany. The country also has a moderate oil production, with fields to the north-east of Vienna and south-west of Linz, and the total output of 2 million tonnes meets one quarter of home demand. There are small deposits of coal, mainly in the south-east, but almost all of it consists of brown coal.

Industry has developed rapidly since the war, especially the iron and steel industry which is located in the Mur-Murz valley and at Linz. Steel production is now 4·5 million tonnes per annum, and is based on both local and imported iron ore, and imported coal and coke. Engineering industries are concerned mainly with railway rolling stock, motor vehicles, machinery and precision instruments. Other leading industries are heavy chemicals, textiles, and the smelting and refining of non-ferrous metals.

Although Austria is a member of the E.F.T.A., most of its trade is with the Common Market countries, especially West Germany and Italy. There is considerable trade with Britain and the United States, and a small but growing trade with the east European countries. The value of the imports exceeds that of the exports, but the adverse balance is largely offset by receipts from foreign tourists since tourism remains a major feature of the Austrian economy.

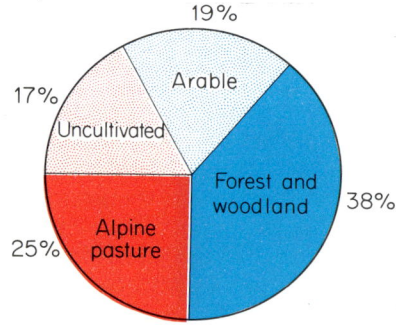

FIG. 12.1 **Land use of Austria**

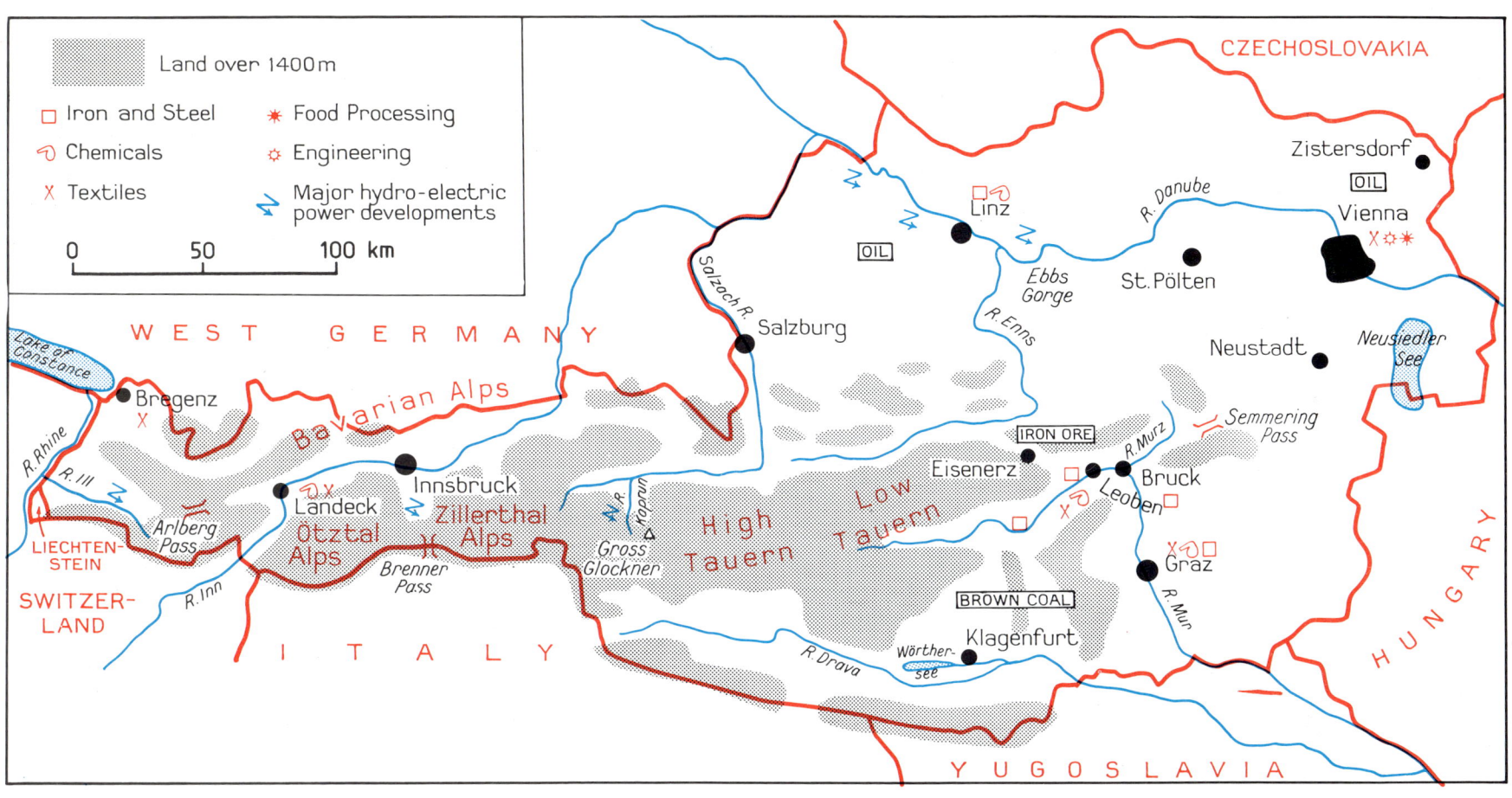

FIG. 12.2 **Austria**

# THE REGIONS OF AUSTRIA

## The Vorarlberg and Tyrol

These western provinces form a narrow region of mountains and valleys. The Vorarlberg extends to the Rhine and Lake Constance, and also to the tiny country of Liechtenstein, which is wedged between Austria and Switzerland. The Tyrol lies to the east of the Arlberg Pass and includes the deep, longitudinal valley of the Inn, with the limestone range of the Bavarian Alps to the north and the rugged crystalline Otztal and Zillerthal Alps to the south. This is Austria's most popular tourist area, and there are numerous resorts catering for the summer visitor as well as the winter sports enthusiast.

The population is concentrated in the valleys, which are well-farmed with fields of hay, oats and roots. Cattle are kept in large numbers and many of them spend the summer on the high Alpine pastures, much of the milk being made into butter and cheese. There are numerous hydro-electric stations, including the new Ill valley scheme in Vorarlberg, and the availability of cheap power has encouraged the growth of industry. Of particular importance are the textile and embroidery industries in Vorarlberg in such towns as *Bregenz* and *Landeck*. The latter also has a chemical industry.

*Innsbruck* (115,000) is the largest town of the region. It is a tourist centre and lies on the swift-flowing Inn, at a bridging point where the east-west route following the river crosses the north-south route from Munich to northern Italy via the Brenner Pass. The latter is the easiest pass of the whole Alpine range, its summit being only 1,375 metres above sea level.

◄ **A valley in the Austrian Alps**     *Austrian State Tourist Department*
Describe the probable occupations of the inhabitants of the village, based on the information in the photograph.

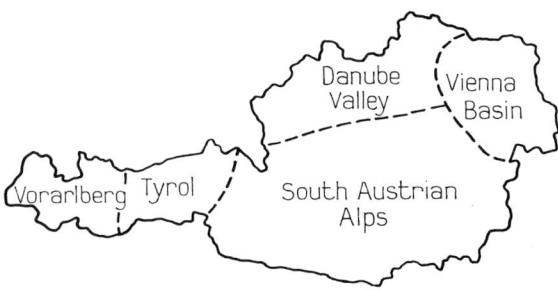

FIG. 12.3 **The regions of Austria**

**Alpine pastures above the treeline** *Austrian State Tourist Department*

## The Southern Austrian Alps

From the Brenner the main crystalline Alps continue east as the High Tauern and Low Tauern, the former rising well above the snow line to rugged, glaciated peaks including the Gross Glockner (3,797 metres), Austria's highest mountain. North of the Tauern ranges are a series of limestone ridges, with spectacular gorges, caverns and lakes, becoming lower as they approach the Danube, whilst to the south another limestone range follows the Italian and Yugoslav frontiers.

The region is one of the country's main sources of hydro-electricity and among the new works completed recently is the Glockner-Kaprun scheme based on water impounded behind several large dams in the Kaprun valley.

The valleys support considerable populations, some of whom are engaged in farming and forestry, some in catering for tourists, and some in manufacturing industries. The Mur-Murz valley, from which Vienna can be reached by the low Semmering Pass, is one of the most important industrial areas in Austria. There are large iron ore workings not far away at *Eisenerz*, and this has led to the development of iron and steel and engineering industries at *Loeben, Bruck* and other towns, using coal brought in from the Ruhr. Farther south is the Drava valley and the fertile Klagenfurt basin, where the warmer summers and shelter from cold winds permit a greater variety of farming activities including the growing of cereals, grape vines and fruit trees, as well as livestock farming. *Klagenfurt* is a focal point of routes from Vienna, Salzburg, Italy and Yugoslavia. *Salzburg* (130,000), is an attractive, historic town lying on the River Salzach near where it leaves the mountains. Formerly it was the centre of an important salt trade, and salt is still mined in the vicinity. Salzburg was the home of Mozart, and its annual Festival of music and opera attracts an ever-increasing number of visitors.

In the eastern part of the region the land slopes towards the Hungarian Plain. The landscape here is monotonous and very like the grassy plains of Hungary, and the climate is more continental, with extremes of temperature and low rainfall. Much of the land is used for **grain-growing and stock rearing**. *Graz* (250,000) lies on the Mur, where it leaves the hills, and on the route from Vienna via the Semmering Pass to Yugoslavia. Local supplies of brown coal and timber, and iron from the upper Mur valley, have encouraged the growth of metal-working, chemical, textile and paper industries.

121

## The Danube Valley and Vienna Basin

The Danube enters Austria from Germany and flows east, via Linz, to Vienna. For much of the way it is bordered by broad water meadows which are liable to flooding after the spring thaw and for this reason are avoided by settlements and main roads. The river however is much used by steamers and trains of barges which travel upstream into Germany and downstream into Czechoslovakia and Hungary. Navigation is hindered by the speed of the current and by winter freezing, but the construction of dams for hydro-electric power above Linz and at the Ebbs gorge has helped to regulate the flow of the river. Both to the north and south of the river the land rises to wooded hills, considerable parts of which have been cleared for pasture and crop-growing. *Linz* (200,000) is an important Danube port and a centre of heavy industry, with large iron and steel works and a chemical industry producing especially fertilisers and pharmaceutical goods. Linz receives power from hydro-electric stations on the Enns and Danube.

The Vienna Basin is a low-lying region between the eastern spurs of the Alps and the Carpathian foothills. It is floored by young sedimentary rocks which have a superficial covering of loess and alluvium, and in the south, on the Hungarian frontier, is the shallow Neusiedler See. The Vienna Basin is Austria's most important farming region. The fertile soils and warm summers make possible the cultivation of wheat, maize, sugar beet and potatoes, and there are numerous vineyards and orchards especially on south-facing slopes. To the north-east of Vienna is the country's principal oilfield near *Zistersdorf*, the oil being taken by pipeline to two refineries on the Danube below Vienna.

*Vienna* (Wien) (1,750,000), Austria's capital and the former imperial capital, occupies rising ground on the south bank of the Danube. The city has spread westwards onto the hilly country known as the Wienerwald (Vienna Woods) and along the Wien valley. It lies at the crossing point of two important routes – one following the Danube valley, and the other from Poland via the Moravian Gate to Yugoslavia and northern Italy.

◄ **The Erzberg iron ore hill near Eisenerz**   *Austrian State Tourist Office*
The Erzberg forms a peak nearly 1,500 m high. The ore is being worked on thirty graded banks each 25 metres high and there are 21 km of mining front here.

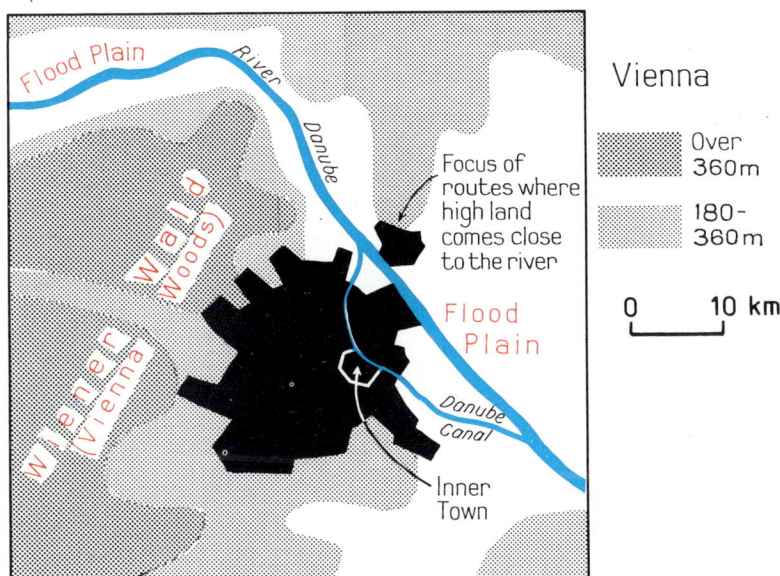

FIG. 12.4 **Vienna**

Vienna is by far the most important commercial and manufacturing centre of Austria. Its manufactures include traditional luxury articles such as high quality clothing, glassware and furniture, and there are newer engineering and electrical industries. The latter have developed mainly in industrial suburbs to the south of the city, and in the area between the Danube and its western branch which is known as the Danube Canal.

The city retains much of its former imperial magnificence, and its many historic buildings, museums and other cultural attractions bring large numbers of foreign tourists.

◄ **The Danube lowland near Vienna**    *Austrian State Tourist Department*
The river here has low, marshy banks, bordered by gently rising ground where there are vineyards and orchards.

123

**The United Austrian Iron and Steel Works at Linz**      *Vereinigte Osterreichische Eisen und Stahlwerke Aktiengesellschaft Linz-Donau*
This is the most modern iron and steel works in Austria. The works are fully integrated, with coke ovens, blast furnaces, steel furnaces and rolling mills and the river can be used for bringing in raw materials.

## Exercises

*Answer in note form :*

1. What sources of power does Austria possess and where are they located?
2. Why is the Vienna Basin the most important farming region in Austria?
3. What part is played by the Danube in the geography of Austria?
4. Why has Austria a large tourist industry?

*Essay Question*

Write an account, illustrated by a sketch map, of the manufacturing industries of Austria.

# INDEX